Marc Kesseböhmer, Sara Munday, Bernd Otto Stratmann
Infinite Ergodic Theory of Numbers
De Gruyter Graduate

Also of Interest

Ergodic Theory
Idris Assani (Ed.), 2016
ISBN 978-3-11-046086-5, e-ISBN (PDF) 978-3-11-046151-0,
e-ISBN (EPUB) 978-3-11-046091-9

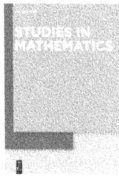

De Gruyter Studies in Mathematics
Carsten Carstensen et al. (Ed.)
ISSN 0179-0986

Volume 62
Positive Dynamical Systems in Discrete Time: Theory, Models, and
Applications
Ulrich Krause, 2015
ISBN 978-3-11-036975-5, e-ISBN (PDF) 978-3-11-036569-6,
e-ISBN (EPUB) 978-3-11-039134-3

Volume 61
Markov Operators, Positive Semigroups and Approximation
Processes
Francesco Altomare, Mirella Cappelletti, Vita Leonessa,
Ioan Rasa, 2014
ISBN 978-3-11-037274-8, e-ISBN (PDF) 978-3-11-036697-6,
e-ISBN (EPUB) 978-3-11-038641-7

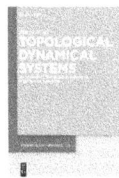

Volume 59
Topological Dynamical Systems: An Introduction to the Dynamics
of Continuous Mappings
Jan Vries, 2014
ISBN 978-3-11-034073-0, e-ISBN (PDF) 978-3-11-034240-6,
e-ISBN (EPUB) 978-3-11-037459-9

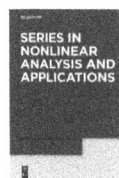

De Gruyter Series in Nonlinear Analysis and Applications
Jürgen Appell et al. (Ed.)
ISSN 0941-813X

Marc Kesseböhmer, Sara Munday,
Bernd Otto Stratmann

Infinite Ergodic Theory of Numbers

—

DE GRUYTER

Mathematics Subject Classification 2010
37Axx, 11B57, 11J70, 11J83, 60K05

Authors
Dr. Sara Munday
Università di Bologna
Department of Mathematics

Professor Dr. Marc Kesseböhmer
Universität Bremen
Department of Mathematics

Professor Dr. Bernd O. Stratmann[†]

ISBN 978-3-11-043941-0
e-ISBN (PDF) 978-3-11-043942-7
e-ISBN (EPUB) 978-3-11-043085-1

Library of Congress Cataloging-in-Publication Data
A CIP catalog record for this book has been applied for at the Library of Congress.

Bibliographic information published by the Deutsche Nationalbibliothek
The Deutsche Nationalbibliothek lists this publication in the Deutsche Nationalbibliografie;
detailed bibliographic data are available on the Internet at http://dnb.dnb.de.

© 2016 Walter de Gruyter GmbH, Berlin/Boston
Typesetting: Konvertus, Haarlem, NL
Printing and binding: CPI books GmbH, Leck
♾ Printed on acid-free paper
Printed in Germany

www.degruyter.com

Preface

This book is intended to serve as an introduction to *Infinite Ergodic Theory* for advanced undergraduate and PhD students and should be appropriate as a text for a seminar or reading course. We hope that some aspects of the presented material will be of interest also to researchers. The prerequisites we have assumed are a certain familiarity with measure theory and (to a lesser extent) basic concepts from functional analysis. For the rest, we have attempted to be as self-contained as possible.

One of the fundamental objectives of ergodic theory is to investigate dynamical systems from a measure-theoretic perspective. Central to this conception are invariant measures – measures which remain unchanged under the dynamics. In classical ergodic theory these measures are probability measures, whereas the topic of infinite ergodic theory is systems which preserve infinite measures. This seemingly small change engenders radically different results, as we will see. A systematic approach to this part of ergodic theory has been provided in the foundational work of Jon Aaronson [Aar97], which is still a major source for a coherent presentation and also guided us in the preparation of this graduate text. In addition to Aaronson's book, we also found helpful the book by Dajani and Kraaikamp [DK02] and the survey article by Stefano Isola [Iso11], as well as unpublished lecture notes by Omri Sarig, Maximilian Thaler, and Roland Zweimüller.

Our aim with this book is to illuminate various aspects of infinite ergodic theory by using several concrete examples of dynamical systems that are strongly linked to number theory. We will use these examples to analyse some explicit questions (like the asymptotic behaviour of sum-level sets) to illustrate not only the powerful methods from infinite ergodic theory but also the strong connection between infinite ergodic theory and renewal theory.

Another theme in the book is elementary Diophantine approximation. We give some classical results for continued fractions in Chapter 1, and, with the intention of closing a circle of ideas, in Chapter 5 we show how the analysis of the sum-level sets for the continued fraction expansion also gives rise to some further Diophantine-type results. The final application in Chapter 5 is to establish a uniform law for the Stern-Brocot sequence contrasting a famous analogous result for the Farey sequence.

The book is organised with chapters containing general theory sandwiched between chapters devoted more to our examples and to applications. The first chapter consists of a little necessary background material and the introduction of all our main examples. It is followed by a chapter containing some standard results in ergodic theory and the beginnings of the theory for infinite systems. Chapter 3 is then devoted to renewal theory and its application to certain piecewise-linear systems. We return to more general infinite ergodic theory for Chapter 4, and finally, in Chapter 5 we see some applications of this theory to the Gauss and Farey systems.

Unfortunately, our good friend, co-author and colleague Bernd O. Stratmann died before the manuscript of this book was complete – we have tried our best to finalise the book also in his spirit. We would like to extend our heartfelt thanks for helpful conversations and for reading various sections to Kathryn Lorenz, Marco Lenci, Giampaolo Cristadoro, Henna Koivusalo, Andrei Ghenciu, Mauro Artigiani, Tushar Das, Niclas Technau, Arne Mosbach and Konstantin Schäfer.

August 2016,
Marc Kesseböhmer and Sara Munday

Contents

Mathematical symbols

$\forall$	'for all'.
$\exists$	'exists'.
0^n	n consecutive appearances of the symbol 0. 32
$\|\cdot\|$	operator norm. 76
$\|\cdot\|_p$	p-norm for $p \in [1, \infty]$. 137
$\varnothing$	empty set.
$\Longrightarrow$	'implies'.
$\Longleftrightarrow$	'equivalent'.
$\#E$	cardinality of E. 18
$:=$	'equal by definition'.
$\overline{A} \subseteq B$	topological closure of A in $\mathbb{R}$. 26
A^*	set of recurrent points in A. 105
$[a, b]$	closed interval.
$[a, b)$	right open interval.
$(a, b]$	left open interval.
(a, b)	open interval.
$A \triangle B$	symmetric difference of the sets A and B. 70
$\mathring{\alpha}$	Markov partition for L_α. 40
α_D	dyadic partition for L_α. 43
α_H	harmonic partition for L_α. 42
$\mathrm{Aut}(\widehat{\mathbb{C}})$	group of automorphisms acting on $\widehat{\mathbb{C}}$. 182
$\mathcal{B}$	badly approximable numbers. 11
$\mathcal{B}_A$	σ-algebra $\mathcal{B}$ restricted to the set A. 106, 108
$\mathcal{B}_{\alpha,N}$	N-badly α-approximable numbers. 54
$\mathcal{B}_{\alpha,\varphi}$	φ-badly α-approximable numbers. 56
$\mathcal{B}_N$	subset of the set of badly approximable numbers. 11
$\mathcal{B}_n$	n-th member of the Stern–Brocot sequence. 33
$B(s)$	$:= \sum_{n=0}^{\infty} b_n s^n$. 131
$\mathbb{C}$	set of complex numbers.
$C_\alpha(\ell_1, \ldots, \ell_k)$	α-Lüroth cylinder set. 44
$\mathcal{C}_b(\mathbb{R})$	real-valued bounded continuous functions. 177
$\mathcal{C}_n$	sum-level set for the continued fraction expansion. 122

C_T	conservative part with respect to the measure-preserving transformation T. 72
C_V	conservative part with respect to the contracting operator V. 141
$C(x_1, \dots, x_n)$	Gauss cylinder. 19
$\widehat{C}(x_1, \dots, x_n)$	Farey cylinder. 33
$\widehat{C}_\alpha(x_1, \dots, x_n)$	α-Farey cylinder. 48
$\widehat{C}_T$	conservative part with respect to $\widehat{T}$ for a non-singular transformation T. 78
Δ_μ	distribution function of a measure μ with support in $[0, 1]$. 36
δ_x	Dirac measure in x. 177
$\mathrm{diam}(A)$	Euclidian diameter of the set A. 28
$\dim_H(A)$	Hausdorff dimension of the set A. 55
$\mathrm{d}\nu/\mathrm{d}\mu$	Radon–Nikodým derivative of ν with respect to μ. 76
$\mathrm{d}(\omega, \tau)$	distance between ω and τ with respect to the metric d. 17
D_T	dissipative part with respect to the measure-preserving transformation T. 72, 78, 141
d_v	greatest common divisor of $\{n > 0 : v_n > 0\}$. 125
$\mathcal{E}$	set of exceptional points of a countable Markov partition. 26
E^*	set of all finite non-empty words over the alphabet E. 17
$E^{\mathbb{N}}$	set of all infinite non-empty words over the alphabet E. 17
ε	empty word; word having no letters. 17
F	Farey map given by $x \mapsto x/(1-x)$ on $[0, 1/2]$, $x \mapsto (1-x)/x$ on $(1/2, 1]$. 30
F^*	jump transformation on $(1/2, 1]$ of the Farey map F. 31
$f^A(x)$	$:= \sum_{j=0}^{\varphi(x)-1} f \circ T^j(x)$; induced version of f on A. 112
F_α	α-Farey map. 44
F_α^*	jump transformation on A_1 of F_α. 46
$F_{\alpha,j}$	inverse branches of the α-Farey map F_α, $j = 0, 1$. 45
F_j	inverse branches of F, $j = 0, 1$. 31
$\mathcal{F}_n$	n-th Farey sequence. 58
$\mathcal{F}^+$	subset of non-negative functions from the function space $\mathcal{F}$. 66
G	Gauss map given by $x \mapsto 1/x - \lfloor 1/x \rfloor$. 15
$\gcd$	greatest common divisor. 125
$\mathrm{GL}_2(\mathbb{Z})$	general linear group. 179
G_n	n-th inverse branch of G. 15
h_F	invariant density of ν_F with respect to λ. 82
h_{F_α}	invariant density of ν_{F_α} with respect to λ. 83

h_G	invariant density of m_G with respect to λ. 85
h_{L_α}	invariant density of m_α with respect to λ. 85
$\mathbb{I}$	set of irrational numbers.
$\mathbb{I}_\alpha$	set of α-irrational numbers. 42
$\mathrm{Int}(A)$	interior of the set A. 40
$\mathfrak{J}_i$	$:= d(\lambda \circ T_i)/d\lambda$; Jacobian for the i-th inverse branch of the interval map T. 81
$\kappa_\pm$	Hölder and sub-Hölder exponent. 49
$L_1(\mu)$	space of μ-integrable functions. 66, 137
$[\ell_1, \ell_2, \ldots, \ell_k]_\alpha$	α-Lüroth expansion. 42
$[\ell_1, \ell_2, \ell_3, \ldots]_\alpha$	α-Lüroth coding. 47
L_α	α-Lüroth map. 40
$L_{\alpha,n}$	inverse branches of L_α. 67
λ	Lebesgue measure. 67
$L_\infty(\mu)$	space of μ-essentially bounded functions. 66, 137
$L(x)$	Lüroth map. 60
$\mathbb{M}$	Markov partition. 27, 40
$m|_A$	measure m restricted to $\mathcal{B}_A$. 106
$M(\mathcal{B})$	equivalence classes of $\mathcal{B}$-measurable functions. 65
$\mathcal{M}(\mathcal{B})$	set of $\mathcal{B}$-measurable functions. 65
m_G	Gauss measure; invariant λ-absolutely continuous measure for G. 67
μ_g	measure with density g with respect to the measure μ. 75, 140
$\mathbb{N}$	set of natural numbers.
$\mathbb{N}_0$	set of natural numbers including 0.
$\nu(x)$	Hurwitz constant for x. 10
ν_F	invariant λ-absolutely continuous measure for F. 82
ν_{F_α}	invariant λ-absolutely continuous measure for F_α. 83
$\nu \ll \mu$	ν absolutely continuous with respect to μ. 75
$\mathcal{O}(x)$	forward orbit of x. 12
$\omega\|n$	initial block $(\omega_1, \omega_2, \ldots, \omega_n)$ of length n of $\omega \in E^*$. 17
$|\omega|$	length of a word ω. 17
$[\omega]$	cylinder set with respect to the finite word $\omega \in E^*$. 17
$\omega \wedge \tau$	longest common initial block of two words ω and τ. 17

$p(x)$	first passage time of x. 86
P_F	Ruelle operator for the map F. 82
P_{F_α}	Ruelle operator for the map F_α. 84
P_G	Ruelle operator for the map G. 85
φ	return time to the set $A \in \mathcal{B}$. 105
φ_n	$:= S_n^A(\varphi)$. 112
p_k/q_k	k-th convergent to the continued fraction. 2
P_T	Ruelle operator for the differential Markov interval map T. 81
$\mathbb{Q}$	set of rational numbers.
$\mathbb{Q}_\alpha$	set of α-rational numbers. 42
Q	Minkowski's question-mark function. 35
$\mathbb{R}$	set of real numbers.
$\rho(x)$	hitting time under F of x to the interval $(1/2, 1]$. 31
$\rho_\alpha(x)$	hitting time under F_α of x to the interval A_1. 45
$r_k^{(\alpha)}(x)$	$:= [\ell_1, \dots, \ell_k]_\alpha$; α-Lüroth convergent of x. 44
r_n	n-th remainder. 4
s_n	$:= q_{n-1}/q_n$. 4
σ	shift map of $E^{\mathbb{N}}$. 18
S_n	$= \mathcal{B}_n \setminus \mathcal{B}_{n-1}$; vertices of the Stern–Brocot Tree. 34
$S_n^A(h)$	ergodic sums for the induced system with respect to $h : A \to \mathbb{R}$. 112
T_E^*	jump transformation on T with respect to E. 86
$\theta_\alpha(x)$	conjugating homeomorphism between $([0, 1], F_\alpha)$ and $([0, 1], T)$. 49
T_i	measurable inverse branches for an interval map T. 80
T_k	denotes the map $x \mapsto kx \mod 1$, $k > 2$, on $\mathbb{R}/\mathbb{Z}$. 14
$\widehat{T}_\mu = \widehat{T}$	transfer operator of T with respect to the measure μ. 76
$\mathcal{T}_n$	Stern–Brocot intervals of order n. 33
$T : X \to Y$	a map from X to Y.
$\mathcal{U} \vee \mathcal{V}$	join of two collections of subsets of X. 27
$\mathcal{U}_\varepsilon(x)$	open neighbourhood of x with diameter ε. 179
$\mathcal{U}^n$	n-th refinement of a collection $\mathcal{U}$. 27
V^*	dual operator of V. 138
$\mathrm{var}_{[a,b]} f$	variation of f over the interval $[a, b]$. 162
$\mathrm{var} f$	total variation of f on $[0, 1]$. 162
V_n	$:= \sum_{k=n}^{\infty} v_k$. 129
$((v_n)_{n \geq 1}, (w_n)_{n \geq 0})$	renewal pair. 124

$\mathcal{W}_\alpha$ well α-approximable numbers. 55

$\mathcal{W}_{\alpha,\varphi}$ φ-well α-approximable numbers. 56

w-$\lim_n \mu_n$ weak-$\star$ limit of the sequence of measures (μ_n). 177

W_T collection of all wandering sets for a map T. 69

X non-empty metric space.

X^* dual space; set of all continuous linear functionals $f : X \to \mathbb{R}$. 77

$\langle x_1, x_2, \ldots \rangle_\alpha$ α-Farey coding. 47

$[x_1, x_2, x_3, \ldots]$ regular continued fraction expansion of an irrational number. 1

$\langle x_1, x_2, x_3, \ldots \rangle$ Farey coding. 32

$[x_1, x_2, \ldots, x_n]$ regular continued fraction expansion of a rational number. 2

$\overline{[x_1, \ldots, x_k]}$ periodic continued fraction expansion. 16

$(X, \mathcal{B}, \mu, T)$ measure-theoretic dynamical system. 64

$\lfloor x \rfloor$ the greatest integer not exceeding x. 5

$\{x\}$ fractional part of x. 5

$(X, \mathcal{B}, \mu)$ σ-finite measure space. 64

(X, T) (topological) dynamical system. 1

$\mathbb{Z}$ set of integers.

1 Number-theoretical dynamical systems

Throughout this book, a *(topological) dynamical system* (X, T) means simply a non-empty metric space X and a continuous map $T : X \to X$. In this chapter, we introduce various dynamical systems that generate real number expansions. Such systems will be referred to as *number-theoretic dynamical systems*. The examples we will consider here are mainly constructed over the unit interval. For further simple examples of number-theoretic dynamical systems, including that of the map that gives rise to the familiar decimal expansion, we refer to Dajani and Kraaikamp [DK02].

1.1 Continued fractions and Diophantine approximation

In this chapter, the main goal is to introduce two well-studied number-theoretic dynamical systems, namely, the Gauss map and the Farey map. Both of these maps are related to the *continued fraction expansion*, which we introduce below. We give here a very succinct introduction, but for more details there are several good books available, for instance the classical text by Khintchine [Khi64] or the more modern approach given by Rockett and Szűsz [RS92].

1.1.1 Continued fractions

An expression of the form

$$[x_1, x_2, x_3, \ldots] := \cfrac{1}{x_1 + \cfrac{1}{x_2 + \cfrac{1}{x_3 + \cdots}}}, \tag{1.1}$$

where each x_i, for $i \in \mathbb{N}$, is a positive integer, is called a *regular continued fraction expansion* (or simply a *continued fraction*). We will refer to the numbers $x_1, x_2, x_3, \ldots$ as the *elements* of the continued fraction. The number of elements of the continued fraction may be finite or infinite. In the first case, we will say we have a *finite* continued fraction and in the second case we will say that we have an *infinite* continued fraction. A finite continued fraction is the result of a finite number of rational operations and hence it represents the rational number given by

$$[x_1, \ldots, x_n] := \cfrac{1}{x_1 + \cfrac{1}{x_2 + \cdots + \cfrac{1}{x_n}}}. \tag{1.2}$$

Notice that for $x_n \geq 2$, the continued fractions $[x_1, \ldots, x_n - 1, 1]$ and $[x_1, \ldots, x_n]$ represent the same number. It is typical to use only the latter expression, but on occasion we find it helpful to have the option of using either.

We cannot immediately assign a value to an infinite continued fraction, so, for the time being, it should be thought of only as a formal notation, akin to that for an infinite series.

In the theory of continued fractions, a particularly important role is played by the initial segments of each (finite or infinite) continued fraction. For a given continued fraction $[x_1, x_2, x_3, \ldots]$, we consider the sequence of rational numbers $([x_1, \ldots, x_k])_{k \geq 1}$. For each $k \in \mathbb{N}$, we will write $p_k/q_k := [x_1, \ldots, x_k]$, where the positive integers p_k and q_k are required to be coprime. Then p_k/q_k will be called the *k-th convergent* to the continued fraction $[x_1, x_2, x_3, \ldots]$. In particular,

$$\frac{p_1}{q_1} = \frac{1}{x_1} \quad \text{and} \quad \frac{p_2}{q_2} = \frac{1}{x_1 + 1/x_2} = \frac{x_2}{x_1 x_2 + 1}.$$

For a finite continued fraction $[x_1, \ldots, x_n]$ with n elements, we have

$$\frac{p_n}{q_n} = [x_1, \ldots, x_n].$$

In this case, there are only n convergents. Each infinite continued fraction has an infinite sequence of convergents. The justification for this terminology will come in Proposition 1.1.3, but first let us give the recurrence relations that describe the formation of the convergents. Here, we make the further definition that $p_0 := 0$ and $q_0 := 1$.

Theorem 1.1.1. *For each $n \in \mathbb{N}$, we have that*
(a) $p_{n+1} = x_{n+1} p_n + p_{n-1}$,
(b) $q_{n+1} = x_{n+1} q_n + q_{n-1}$,
(c) $q_n p_{n-1} - p_n q_{n-1} = (-1)^n$.

Proof. The statements in (a) and (b) can be proved by induction; we leave this as an exercise for the reader. For part (c), multiplying part (a) by q_n and part (b) by p_n, then subtracting the first from the second yields

$$q_{n+1} p_n - p_{n+1} q_n = -(q_n p_{n-1} - p_n q_{n-1}).$$

It then suffices to notice that $q_1 p_0 - p_1 q_0 = -1$ to complete the proof of the theorem. □

Part (c) of Theorem 1.1.1 has the following immediate and useful corollary.

Corollary 1.1.2.
(a) *For all $n \geq 1$,*

$$\frac{p_{n-1}}{q_{n-1}} - \frac{p_n}{q_n} = \frac{(-1)^n}{q_n q_{n-1}}.$$

(b) *For all $n \geq 2$,*

$$\frac{p_{n-2}}{q_{n-2}} - \frac{p_n}{q_n} = \frac{(-1)^{n-1} x_n}{q_n q_{n-2}}.$$

This corollary allows us to reach important conclusions about the sequence of convergents to a continued fraction. Specifically, we have the following result.

Proposition 1.1.3. *The sequence of convergents $(p_n/q_n)_{n \geq 1}$ of the continued fraction $[x_1, x_2, x_3, \ldots]$ satisfies the following four properties:*
(a) *The sequence $(p_{2n}/q_{2n})_{n \geq 1}$ of even convergents is increasing.*
(b) *The sequence $(p_{2n-1}/q_{2n-1})_{n \geq 1}$ of odd convergents is decreasing.*
(c) *Every convergent of odd order is greater than every convergent of even order.*
(d) $\lim\limits_{n \to \infty} \left(\dfrac{p_{n+1}}{q_{n+1}} - \dfrac{p_n}{q_n} \right) = 0.$

Proof. The statements in parts (a) and (b) follow directly from Corollary 1.1.2 (b). For part (c), observe that by Corollary 1.1.2 (a) we have that every odd convergent is greater than the directly preceding even convergent. Thus, p_{2k+1}/q_{2k+1} is greater than all of the smaller-index even convergents: $p_2/q_2, p_4/q_4, \ldots, p_{2k}/q_{2k}$. Suppose that there exists some convergent p_{2m}/q_{2m}, where $m > k$, such that

$$\frac{p_{2m}}{q_{2m}} > \frac{p_{2k+1}}{q_{2k+1}}.$$

But then,

$$\frac{p_{2m}}{q_{2m}} < \frac{p_{2m+1}}{q_{2m+1}} < \frac{p_{2k+1}}{q_{2k+1}},$$

where the last inequality comes again from Corollary 1.1.2 (a). This contradiction finishes the proof of part (c). Finally, for part (d), it suffices to show that

$$q_n \geq 2^{\frac{n-1}{2}}. \tag{1.3}$$

Indeed, directly from Theorem 1.1.1 (b), we have that

$$q_n = x_n q_{n-1} + q_{n-2} \geq q_{n-1} + q_{n-2} \geq 2 q_{n-2}.$$

Thus, repeated applications of the above inequality yields

$$q_{2n} \geq 2^n q_0 = 2^n \text{ and } q_{2n+1} \geq 2^n q_1 \geq 2^n.$$

This finishes the proof. □

Using the above proposition, it then makes sense to say that the value of the infinite continued fraction $[x_1, x_2, x_3, \ldots]$ is equal to the real number that is the limit of the sequence of convergents associated with it.

Definition 1.1.4. For $x = [x_1, x_2, x_3, \ldots] \in (0, 1]$ and $n \in \mathbb{N}$, let r_n and s_n be defined as follows:

$$r_n := \frac{1}{[x_n, x_{n+1}, x_{n+2}, \ldots]} \text{ and } s_n := \frac{q_{n-1}}{q_n}.$$

The number r_n is called the *n-th remainder* of x.

Note that for $x = [x_1, x_2, x_3, \ldots]$ and $n \in \mathbb{N}$, the following recurrence relation holds:

$$r_n = x_n + \frac{1}{r_{n+1}}.$$

Since $q_n = x_n q_{n-1} + q_{n-2}$, we have that

$$s_n^{-1} = \frac{q_n}{q_{n-1}} = x_n + \frac{1}{q_{n-1}/q_{n-2}}.$$

Clearly, this process may be continued until $q_1/q_0 = x_1$ is reached. Therefore,

$$s_n = [x_n, x_{n-1}, \ldots, x_1].$$

Theorem 1.1.5. *For $x = [x_1, x_2, x_3, \ldots] \in (0, 1]$ and $n \in \mathbb{N}$, we have that*

$$x = \frac{p_n r_{n+1} + p_{n-1}}{q_n r_{n+1} + q_{n-1}}.$$

Proof. We will prove the theorem by induction. For $n = 1$, on recalling the definitions $p_0 := 0$ and $q_0 := 1$, we deduce that

$$\frac{p_1 r_2 + p_0}{q_1 r_2 + q_0} = \frac{r_2 + 0}{x_1 r_2 + 1} = \frac{1}{x_1 + 1/r_2} = x.$$

Now assume that the statement is true for some $n \in \mathbb{N}$. Then, in light of Theorem 1.1.1 (c), we obtain that

$$x = \frac{p_n r_{n+1} + p_{n-1}}{q_n r_{n+1} + q_{n-1}} = \frac{p_n(x_{n+1} + 1/r_{n+2}) + p_{n-1}}{q_n(x_{n+1} + 1/r_{n+2}) + q_{n-1}}$$
$$= \frac{p_n x_{n+1} r_{n+2} + p_n + p_{n-1} r_{n+2}}{q_n x_{n+1} r_{n+2} + q_n + q_{n-1} r_{n+2}}$$
$$= \frac{p_{n+1} r_{n+2} + p_n}{q_{n+1} r_{n+2} + q_n}.$$

□

This theorem allows us to calculate the distance between a real number $x = [x_1, x_2, x_3, \ldots]$ and any of its convergents.

Corollary 1.1.6. *For $x = [x_1, x_2, x_3, \ldots] \in (0, 1]$ and $n \in \mathbb{N}$, we have that*

$$\left| x - \frac{p_n}{q_n} \right| = \frac{1}{q_n^2(r_{n+1} + s_n)} < \frac{1}{x_{n+1}q_n^2} \le \frac{1}{q_n^2}.$$

Proof. Using Theorem 1.1.5 and Theorem 1.1.1 (c), we infer that

$$\left| x - \frac{p_n}{q_n} \right| = \left| \frac{p_n r_{n+1} + p_{n-1}}{q_n r_{n+1} + q_{n-1}} - \frac{p_n}{q_n} \right|$$

$$= \left| \frac{p_n q_n r_{n+1} + p_{n-1} q_n - p_n q_n r_{n+1} - p_n q_{n-1}}{(q_n r_{n+1} + q_{n-1}) q_n} \right|$$

$$= \left| \frac{q_n p_{n-1} - p_n q_{n-1}}{q_n^2(r_{n+1} + s_n)} \right| = \frac{1}{q_n^2(r_{n+1} + s_n)}.$$

The first inequality follows since $x_{n+1} < r_{n+1}$, the second since $x_{n+1} \ge 1$. □

We end this section with the important result that every real number admits a continued fraction expansion. The basis of the proof is simply the Euclidean algorithm, the algorithm for finding the greatest common divisor of two integers. Before stating the theorem, recall that for each positive real number x the notation $\lfloor x \rfloor$ denotes the greatest integer not exceeding x and $\{x\}$ denotes the fractional part of x, that is, $x = \lfloor x \rfloor + \{x\}$.

Theorem 1.1.7. *To every real number $x \in (0, 1]$ there corresponds a continued fraction with value equal to x. This continued fraction is infinite if and only if x is irrational. Moreover, every irrational number has a unique continued fraction expansion.*

Proof. If $x = 1$, then the continued fraction $[1] := 1/1$ is the one sought. So, suppose that $x \in (0, 1)$. Then set $r_1 := 1/x$ and define $x_1 := \lfloor r_1 \rfloor$, so that

$$x = \frac{1}{x_1 + \{r_1\}}.$$

If r_1 is an integer, we are finished. Otherwise, if $\{r_1\} \in (0, 1)$, then we set $r_2 := 1/\{r_1\}$, so that

$$x = \frac{1}{x_1 + 1/r_2}.$$

Suppose that the numbers $r_1, r_2, \ldots, r_n$ have been defined and, if r_n is not an integer, let $x_n := \lfloor r_n \rfloor$ and set $r_{n+1} := 1/\{r_n\}$. Then we obtain the relation

$$r_n = x_n + \frac{1}{r_{n+1}}.$$

If the number x happens to be a rational number, then each r_n as defined above will also be rational. In this case, the process must stop after a finite number of steps. Indeed, if $r_n = a/b$ and r_n is not already an integer, then

$$r_{n+1} = \frac{1}{r_n - x_n} = \frac{1}{a/b - x_n} = \frac{b}{a - x_n b}.$$

Therefore, r_{n+1} has a smaller denominator than r_n and it follows that if we consider the sequence $r_1, r_2, r_3, \ldots$, we must eventually come to an integer. If that integer is r_k, then the number x is represented by the finite continued fraction $[x_1, x_2, \ldots, x_k]$, where $x_k := r_k > 1$ (if $r_k = 1$, then r_{k-1} must also be an integer, so we replace the two final terms x_{k-1} and 1 by the single integer $x_{k-1} + 1$).

If x is irrational, then each r_n must also be irrational and the above-described process will not terminate. Then, by the definition of x_k and Proposition 1.1.3, where $p_n/q_n := [x_1, x_2, \ldots, x_n]$ as before, we have for each $n \geq 1$ that

$$\frac{p_{2n}}{q_{2n}} < x < \frac{p_{2n+1}}{q_{2n+1}}.$$

Thus, it follows by Proposition 1.1.3 (d) that

$$\lim_{n \to \infty} \frac{p_n}{q_n} = x.$$

This means that the continued fraction $[x_1, x_2, x_3, \ldots]$ has as its value the given irrational number x.

It only remains to show that each infinite expansion is unique (recall that the finite expansions are not unique, since the value of $[x_1, \ldots, x_n]$ is equal to the value of $[x_1, \ldots, x_n - 1, 1]$, for $x_n \geq 2$). So, fix

$$x := [x_1, x_2, x_3, \ldots] \text{ and } x' := [x'_1, x'_2, x'_3, \ldots],$$

and suppose that $n + 1 := \min\{k \in \mathbb{N} : x_k \neq x'_k\}$. Since the first n approximants of x and x' must coincide, but the remainders r_{n+1} of x and r'_{n+1} of x' differ, it follows from Theorem 1.1.5 and the strict monotonicity of $r \mapsto (p_n r + p_{n-1})/(q_n r + q_{n-1})$ that

$$x = \frac{p_n r_{n+1} + p_{n-1}}{q_n r_{n+1} + q_{n-1}} \neq \frac{p_n r'_{n+1} + p_{n-1}}{q_n r'_{n+1} + q_{n-1}} = x'.$$

This proves uniqueness of the continued fraction expansion for all irrational numbers from the unit interval. □

1.1.2 Elementary Diophantine approximation: Hurwitz's Theorem and badly approximable numbers

Let us now turn to some number-theoretic applications of continued fractions. One useful property of the continued fraction expansion of an irrational number is that it allows us to approximate the value of this number by rational numbers to within any desired degree of accuracy. These approximations are given by the convergents (this is why the convergents are sometimes also referred to as *approximants*). The larger $n \in \mathbb{N}$ is chosen, the closer the rational number p_n/q_n comes in value to $x = [x_1, x_2, x_3, \ldots]$. The results presented below all concern the closeness, in absolute value, of the convergents to the irrational number they are approximating. These sorts of questions fall into the area of mathematics known as *Diophantine approximation*. This is a vast and extremely active research area with many interesting and deep open problems.

Theorem 1.1.8. *For all irrational numbers $x = [x_1, x_2, x_3, \ldots]$ and for all $n \in \mathbb{N}$, we have that the inequality*

$$\left| x - \frac{p_i}{q_i} \right| < \frac{1}{2q_i^2}$$

is fulfilled for at least one element $i \in \{n, n+1\}$.

Proof. By way of contradiction, suppose that the statement in the theorem is false. This means that there exists some $n \in \mathbb{N}$ such that the inequality

$$\left| x - \frac{p_i}{q_i} \right| \geq \frac{1}{2q_i^2}$$

holds simultaneously for $i = n$ and $i = n+1$. Since, in light of Corollary 1.1.6, we have that $|x - p_i/q_i| = (q_i^2(r_{i+1} + s_i))^{-1}$, this is equivalent to

$$r_{i+1} + s_i \leq 2, \quad \text{for } i = n, n+1.$$

Let us consider each case separately.

(i) For $i = n$, we can rewrite this as $2 \geq r_{n+1} + s_n = x_{n+1} + 1/r_{n+2} + s_n$ and hence,

$$\frac{1}{r_{n+2}} \leq 2 - (x_{n+1} + s_n) = 2 - \frac{1}{s_{n+1}}.$$

(ii) For $i = n+1$, we obtain that

$$r_{n+2} \leq 2 - s_{n+1}.$$

Combining (i) and (ii), we infer that $1 \le 4 - 2s_{n+1} - 2s_{n+1}^{-1} + 1$. It therefore follows that $0 \le 2 - s_{n+1} - s_{n+1}^{-1}$, which finally implies that

$$0 \ge (s_{n+1} - 1)^2.$$

Since for each $n \in \mathbb{N}$ we have that $s_{n+1} \ne 1$, this contradiction finishes the proof. □

Theorem 1.1.9 (Hurwitz's Theorem I). *For all irrational numbers $x = [x_1, x_2, x_3, \ldots]$ and for all $n \in \mathbb{N}$, we have that the inequality*

$$\left| x - \frac{p_i}{q_i} \right| < \frac{1}{\sqrt{5}\, q_i^2}$$

is fulfilled for at least one element $i \in \{n, n+1, n+2\}$.

Proof. As in the proof of the previous theorem (with 2 replaced by $\sqrt{5}$), assume by way of contradiction that for each $i \in \{n, n+1, n+2\}$ we have that

$$r_{i+1} + s_i \le \sqrt{5}.$$

Proceeding for $i = n$ and $i = n + 1$ as in (i) and (ii) in the previous proof, we derive the inequality

$$s_{n+1}^2 - \sqrt{5}s_{n+1} + 1 \le 0. \tag{1.4}$$

Analogously, by considering in turn $i = n + 1$ and $i = n + 2$, we deduce that

$$s_{n+2}^2 - \sqrt{5}s_{n+2} + 1 \le 0. \tag{1.5}$$

By the quadratic formula, (1.4) and (1.5) yield, where $\gamma := (\sqrt{5} + 1)/2$ and $\gamma^* := (\sqrt{5} - 1)/2$,

$$\gamma^* < s_i < \gamma, \quad \text{for } i = n+1, n+2. \tag{1.6}$$

The strict inequality follows from the fact that γ and γ^* are both irrational and s_n is rational. Using this, we obtain that

$$s_{n+2} = \frac{1}{x_{n+2} + s_{n+1}} \le \frac{1}{1 + s_{n+1}} < \frac{1}{1 + \gamma^*} = \gamma^*,$$

which contradicts (1.6). This finishes the proof. □

Remark 1.1.10. In this context, the number $1/\sqrt{5}$ is called the *Hurwitz number*. Sometimes this number is called the *Hurwitz constant for the golden mean*, and thus there are many others in the sense of Definition 1.1.13 (b).

The number $\gamma := (\sqrt{5} + 1)/2$ which appears in the above proof is known as the *golden mean* (or, sometimes, the *golden ratio*). There are a great many commonly-held beliefs about the golden mean and why it is supposed to be so interesting and/or important; unfortunately, a lot of these are simply wrong (see [Mar92]). However, the continued fraction expansion of γ does give a clue as to why this particular number turns up so often. Observe that γ is one of the two roots of the equation $x^2 - x - 1 = 0$. Writing this another way, we have that

$$\gamma = 1 + \frac{1}{\gamma}.$$

Then, substituting for the γ which appears on the right-hand side of the above equality, we obtain that

$$\gamma = 1 + \cfrac{1}{1 + \cfrac{1}{\gamma}}.$$

Clearly this process can be repeated infinitely often, to yield the continued fraction expansion $\gamma = 1 + [1, 1, 1, \ldots]$, where the fractional part consists of infinitely many ones. The other number γ^* appearing in the proof of Theorem 1.1.9 is simply equal to $\gamma - 1 = [1, 1, 1, \ldots]$. Since we are mostly concerned in this book with numbers in the unit interval, we shall, in a slight abuse of terminology, also refer to this number as the golden mean.

The next theorem shows that the constant $1/\sqrt{5}$ that appears in Theorem 1.1.9 cannot be improved for arbitrary irrational numbers. The proof again relies upon the golden mean.

Theorem 1.1.11 (Hurwitz's Theorem II). *For the golden mean γ^* we have that the inequality*

$$\left| \gamma^* - \frac{p_n}{q_n} \right| \le \frac{C}{q_n^2}$$

is satisfied for at most finitely many convergents p_n/q_n if and only if $C < 1/\sqrt{5}$.

Proof. Firstly, note that since $\gamma^* = [1, 1, 1, \ldots]$, we have for the remainders that $r_n := x_n + [x_{n+1}, x_{n+2}, \ldots]$ is in this case given by $r_n = 1 + [1, 1, 1, \ldots] = 1/\gamma^*$, for all $n \in \mathbb{N}$. Secondly, note that

$$s_n = [x_n, x_{n-1}, \ldots, x_1] = \gamma^* + ([x_n, \ldots, x_1] - [x_{n+1}, x_{n+2}, \ldots]) = \gamma^* + \delta_n,$$

where for δ_n we have that $\lim_{n \to \infty} \delta_n = 0$. Hence, from these two observations it follows that

$$r_{n+1} + s_n = \frac{1}{\gamma^*} + \gamma^* + \delta_n = \frac{\sqrt{5}+1}{2} + \frac{\sqrt{5}-1}{2} + \delta_n = \sqrt{5} + \delta_n.$$

Now, if $C < \frac{1}{\sqrt{5}}$ is given, say $C = \frac{1}{\sqrt{5}+\rho}$ for some fixed $\rho > 0$, then

$$\left| \gamma - \frac{p_n}{q_n} \right| = \frac{1}{q_n^2(r_{n+1}+s_n)} = \frac{1}{q_n^2(\sqrt{5}+\delta_n)} \leq \frac{1}{q_n^2(\sqrt{5}+\rho)},$$

where the last inequality can only be fulfilled for finitely many n, due to the fact that $\sqrt{5}+\rho < \sqrt{5}+\delta_n$ can be satisfied for at most finitely many n. $\qquad\square$

Corollary 1.1.12. *For each irrational number $x \in (0, 1]$, the inequality*

$$\left| x - \frac{p_n}{q_n} \right| \leq \frac{K}{q_n^2}$$

is fulfilled for infinitely many reduced p_n/q_n as long as $K \geq 1/\sqrt{5}$.

We now want to investigate some further results in the vein of Hurwitz's Theorems and Corollary 1.1.12 given above. Before that, it will be useful to make some further definitions.

Definition 1.1.13.

(a) Let c denote a fixed positive real number. An irrational number x is said to be *c-approximable* if and only if the inequality

$$\left| x - \frac{p}{q} \right| < \frac{c}{q^2}$$

is satisfied for infinitely many reduced p/q.

(b) To each irrational number x we associate a non-negative real number $\nu(x)$, defined by

$$\nu(x) := \inf\{c > 0 : x \text{ is } c\text{-approximable}\},$$

and called the *Hurwitz constant* for x.

(c) Two irrational numbers x and y are called *equivalent* if and only if there exist $k, \ell \in \mathbb{N}$ such that the k-th remainder of x is equal to the ℓ-th remainder of y. In other words, the continued fraction expansions of x and y are the same if we discard a finite initial block from each.

(d) Irrational numbers equivalent to the golden mean are called *noble numbers*.

So, for every irrational number x it follows directly from Theorem 1.1.9 (Hurwitz's Theorem I) that $\nu(x) \leq 1/\sqrt{5}$. From Hurwitz's Theorem II and the fact that if x and y are equivalent then $\nu(x) = \nu(y)$ (see Exercise 1.6.6), it follows that if an irrational number x is noble, then $\nu(x) = 1/\sqrt{5}$.

Theorem 1.1.14. *Let N be some fixed positive integer. If $x = [x_1, x_2, x_3, \ldots]$ is irrational such that for some $n \in \mathbb{N}$ we have that the inequality*

$$\left| x - \frac{p_i}{q_i} \right| > \frac{1}{q_i^2 \sqrt{N^2 + 4}}$$

is fulfilled for all $i \in \{n, n+1, n+2\}$, then it follows that $x_{n+2} < N$.

Proof. We proceed as in the proofs of Theorems 1.1.8 and 1.1.9, with 2 and $\sqrt{5}$, respectively, now replaced by $\sqrt{N^2 + 4}$. In this way, considering in sequence $i = n$ and $i = n+1$, we derive

$$s_{n+1}^2 - \sqrt{N^2 + 4}\, s_{n+1} + 1 < 0.$$

Also, by considering $i = n+1$ and $i = n+2$, we similarly derive the inequality

$$s_{n+2}^2 - \sqrt{N^2 + 4}\, s_{n+2} + 1 < 0.$$

Then, using the quadratic formula, we obtain that

$$\frac{\sqrt{N^2 + 4} - N}{2} < s_i \text{ and } s_i^{-1} < \frac{\sqrt{N^2 + 4} + N}{2}, \quad \text{for } i = n+1, n+2.$$

Using this, we then have that

$$x_{n+2} = s_{n+1} + x_{n+2} - s_{n+1} = s_{n+2}^{-1} - s_{n+1}$$
$$< \frac{\sqrt{N^2 + 4} + N}{2} - \frac{\sqrt{N^2 + 4} - N}{2} = N. \qquad \square$$

It follows on combining the above theorem with the second theorem of Hurwitz that if $\nu(x) = 1/\sqrt{5}$, then for all large enough n, we have $x_n = 1$. That is, if $\nu(x) = 1/\sqrt{5}$, then x is a noble number. In light of the discussion immediately preceding Theorem 1.1.14, we have that $\nu(x) = 1/\sqrt{5}$ if and only if x is a noble number.

Let us now consider another interesting class of numbers, namely, all those with bounded continued fraction elements.

Definition 1.1.15. For each $N \in \mathbb{N}$, define

$$\mathcal{B}_N := \{[x_1, x_2, \ldots] \in [0, 1] \setminus \mathbb{Q} : \exists\, n_0 \in \mathbb{N} \text{ such that } x_n < N \text{ for all } n \geq n_0\}.$$

The set of *badly approximable* numbers $\mathcal{B}$ is then defined to be

$$\mathcal{B} := \bigcup_{N > 0} \mathcal{B}_N = \{x \in [0, 1] \setminus \mathbb{Q} : \text{there exists } N \in \mathbb{N} \text{ such that } x \in \mathcal{B}_N\}.$$

In other words, an irrational number x belongs to the set $\mathcal{B}_N$ if and only if it is equivalent to some $y = [y_1, y_2, \ldots]$ such that $y_i < N$ for all $i \in \mathbb{N}$. The following proposition clarifies why the elements in $\mathcal{B}$ are called "badly approximable".

Proposition 1.1.16.
(a) *Fix $N \in \mathbb{N}$. If x is an irrational number in $[0, 1]$ such that $x \notin \mathcal{B}_N$, then*

$$\left| x - \frac{p_n}{q_n} \right| \le \frac{1}{q_n^2 \sqrt{N^2 + 4}}$$

is fulfilled for infinitely many $n \in \mathbb{N}$. In other words, we have that $\nu(x) \le 1/\sqrt{N^2 + 4}$.
(b) *For each $x \in \mathcal{B}$ there exists a constant $C > 0$ such that for all $n \in \mathbb{N}$ we have*

$$\left| x - \frac{p_n}{q_n} \right| > \frac{C}{q_n^2}.$$

Proof. The first part is an immediate consequence of Theorem 1.1.14. For the second part, fix $x = [x_1, x_2, x_3, \ldots] \in \mathcal{B}$. Then there exist numbers M and m_0 such that $x_n < M$ for all $n \ge m_0$. Using this, for such an n we derive the inequality

$$r_{n+1} + s_n = x_{n+1} + [x_{n+2}, x_{n+3}, \ldots] + [x_n, \ldots, x_1] < M + 1 + 1 = M + 2$$

and hence

$$\left| x - \frac{p_n}{q_n} \right| > \frac{1}{q_n^2 (M + 2)}, \quad \text{for all } n \ge m_0.$$

For each of the finitely many $n < m_0$ we have that there exists a number $c_n > 0$ such that

$$\left| x - \frac{p_n}{q_n} \right| > \frac{c_n}{q_n^2}.$$

If we then define $C := \min\{1/(M + 2), c_0, c_1, \ldots, c_{m_0-1}\}$, the result follows. □

Notice that in the above proof, it can be seen that the constant C depends on the value of N for which $x \in \mathcal{B}_N$. In that sense, one might say that the noble numbers are the "worst approximable" numbers.

1.2 Topological Dynamical Systems

Throughout this book, a *(topological) dynamical system* (X, T) is a continuous map $T : X \to X$ of a non-empty metric space X. The first objective in the study of a dynamical system is the consideration of the *orbits* of points of X. The (forward) orbit $\mathcal{O}(x)$ is the set

$$\mathcal{O}(x) := \{T^n(x) : n \ge 0\}.$$

For certain types of point, which we define below, the orbit is very easy to determine.

Definition 1.2.1. Let $T : X \to X$ be a dynamical system.
(a) If $T(x) = x$, then x is called a *fixed point* for T.
(b) A point $x \in X$ is said to be *periodic* for T if $T^n(x) = x$ for some $n \geq 1$. Then n is called a *period* of x. The smallest period of a periodic point x is called the *prime period* of x. Note that the fixed points for T are the periodic points with prime period $n = 1$.
(c) A point $x \in X$ is said to be *pre-periodic* for T, if $T^k(x)$ is a periodic point for some $k \geq 1$. In other words, x is pre-periodic if $T^{k+n}(x) = T^k(x)$ for some $n \geq 1$.

Suppose that we are given two dynamical systems (X, T) and (Y, S). It is desirable to have conditions under which these two systems should be considered dynamically equivalent, that is, as in some dynamical sense, "the same". The sense we are after is that their orbits should behave in the same way. The following definition does exactly this job.

Definition 1.2.2. Two dynamical systems (X, T) and (Y, S) are said to be *topologically conjugate* if there exists a homeomorphism $h : X \to Y$, called a *conjugacy map*, such that

$$h \circ T = S \circ h.$$

In other words, (X, T) and (Y, S) are topologically conjugate if there exists a homeomorphism h such that the following diagram commutes:

$$
\begin{array}{ccc}
X & \xrightarrow{\;T\;} & X \\
\downarrow{\scriptstyle h} & & \downarrow{\scriptstyle h} \\
Y & \xrightarrow{\;S\;} & Y
\end{array}
$$

Remark 1.2.3.
1. Topological conjugacy defines an equivalence relation on the space of all topological dynamical systems.
2. If two dynamical systems (X, T) and (Y, S) are topologically conjugate via a conjugacy map h, then all of their corresponding iterates are topologically conjugate by means of h. That is, $h \circ T^n = S^n \circ h$ for all $n \geq 1$. Therefore, there exists a one-to-one correspondence between the orbits of T and those of S.

Suppose that we are given two topologically conjugate dynamical systems, (X, T) and (Y, S). If $T(x) = x$, it follows that

$$h(x) = h \circ T(x) = S(h(x))$$

and so $h(x)$ is a fixed point of the map S. Thus, there is a one-to-one correspondence between the fixed points of T and the fixed points of S. In particular, if the number of fixed points of T and S are not equal, the systems cannot be topologically conjugate. The number of fixed points is an example of a *topological conjugacy invariant*. The number of periodic points of each prime period is similarly seen to be a topological conjugacy invariant.

Example 1.2.4.
(a) Consider the two maps of the unit circle $T_2 : \mathbb{R}/\mathbb{Z} \to \mathbb{R}/\mathbb{Z}$ and $T_3 : \mathbb{R}/\mathbb{Z} \to \mathbb{R}/\mathbb{Z}$, defined by setting

$$T_2(x) := 2x \,(\mathrm{mod}\ 1) \ \text{ and } \ T_3(x) := 3x \,(\mathrm{mod}\ 1).$$

The map T_2 has a single fixed point in 0 and the set of fixed points of the map T_3 is $\{0, 1/2\}$. Therefore, the number of fixed points is not the same and so, the sytems $(\mathbb{R}/\mathbb{Z}, T_2)$ and $(\mathbb{R}/\mathbb{Z}, T_3)$ cannot be topologically conjugate.
(b) Let $f : [0, 1] \to [0, 1]$ be given by $f(x) := \sqrt{x}$ and $g : [0, 1] \to [0, 1]$ be given by $g(x) := 3x(1 - x)$. Then the set of fixed points of f is $\{0, 1\}$, whereas the set of fixed points of g is $\{0, 2/3\}$. However, although they have the same number of fixed points, there is no topological conjugacy map between $([0, 1], f)$ and $([0, 1], g)$. This can be seen by noting that every homeomorphism $h : [0, 1] \to [0, 1]$ is either strictly increasing or strictly decreasing. In order to have h be a conjugating homeomorphism between f and g, we would have to have $h(0) := 0$ and $h(1) := 2/3$ or vice versa. But this is simply not possible for a strictly monotonic function that also has to map $[0, 1]$ onto $[0, 1]$.

In the next definition, we give a weaker notion than that of topological conjugacy.

Definition 1.2.5. Let (X, T) and (Y, S) be two dynamical systems. If there exists a continuous surjection $h : X \to Y$ which satisfies $h \circ T = S \circ h$, then S is called a (topological) *factor* of T. The map h is thereafter called a *factor map*.

In general, the existence of a factor map between two systems is not sufficient to make them topologically conjugate. Nonetheless, if (Y, S) is a factor of (X, T), then every orbit of T is projected to an orbit of S. As every factor map is by definition surjective, this means that all of the orbits of S have an analogue in T. However, as a factor map may not be injective, more than one orbit of T may be projected to the same orbit of S. In other words, some orbits of S may have more than one analogue in T. Therefore, the dynamical system (X, T) can in this sense be thought of as being more "complicated" than the factor (Y, S).

In the following subsection, we introduce the first of the examples of number-theoretic dynamical systems that will be used to illustrate various concepts throughout the book. We will soon see that this map is related to the continued fraction expansion introduced in Section 1.1.1.

1.2.1 The Gauss map

Definition 1.2.6. Let $G : [0, 1] \to [0, 1]$ be defined by

$$G(x) := \begin{cases} \dfrac{1}{x} - \left\lfloor \dfrac{1}{x} \right\rfloor & \text{for } 0 < x \le 1; \\ 0 & \text{for } x = 0. \end{cases}$$

The map G is referred to as the *Gauss map* and its graph is shown in Fig. 1.1.

Let us also define here the *inverse branches* $G_n : (0, 1) \to (1/(n + 1), 1/n)$ of the Gauss map. These are given, for each $n \in \mathbb{N}$, by

$$G_n(x) := \frac{1}{x + n}.$$

In the following proposition, we shall show how the map G acts on points in the unit interval written in terms of their continued fraction expansion.

Proposition 1.2.7. *If* $x = [x_1, x_2, x_3, \ldots] \in [0, 1]$, *where the continued fraction expansion of* x *is either infinite or finite and consists of at least two elements, then* $G(x) = [x_2, x_3, x_4 \ldots]$. *Moreover, if* $x = [x_1]$, *then* $G(x) = 0$.

Proof. If $x = [x_1, x_2, x_3, \ldots]$, then directly from the definition of G, we have that

$$G(x) = \frac{1}{x} - \left\lfloor \frac{1}{x} \right\rfloor = x_1 + \cfrac{1}{x_2 + \cfrac{1}{x_3 + \ldots}} - x_1 = [x_2, x_3, \ldots].$$

If $x = [x_1] = 1/x_1$, then $G(x) = x_1 - x_1 = 0$. $\qquad\qquad\square$

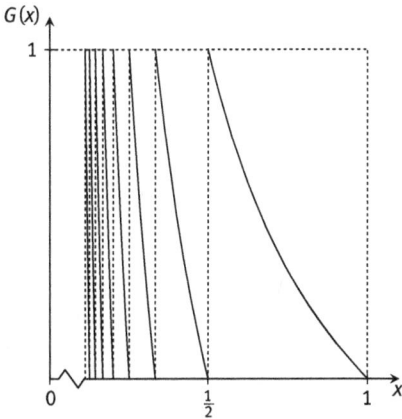

Fig. 1.1. The Gauss map $G : [0, 1] \to [0, 1]$, $G(x) := \dfrac{1}{x} - \left\lfloor \dfrac{1}{x} \right\rfloor$ for $x \in (0, 1]$ and $G(0) := 0$.

Using the latter proposition, we can very easily identify the fixed points and periodic points for the map G. First of all, by definition, G has a fixed point at 0. There are countably many more fixed points, given by the points $x = [n, n, n, \ldots]$, for $n \in \mathbb{N}$. For $n = 1$, this is the golden mean, γ^*. For $n = 2$, we have the fixed point $\sqrt{2} - 1 = [2, 2, 2, \ldots]$. The periodic points for G are simply the periodic continued fractions, that is, points with continued fraction expansions of the form $[\overline{x_1, \ldots, x_k}] :=$ $[x_1, \ldots, x_k, x_1, \ldots, x_k, x_1, \ldots]$, with the block $x_1, \ldots, x_k$ repeating infinitely many times. It remains to describe the pre-periodic points. The pre-periodic points of the trivial fixed point 0 are the rational numbers. The pre-periodic points for all other fixed points are numbers of the form $x = [x_1, \ldots, x_k, n, n, n, \ldots]$, that is, with finitely many elements that can take any value, before infinitely many of the same element n, for some $n \in \mathbb{N}$. In particular, the pre-periodic points for the fixed point γ^* are the set of noble numbers (cf. Definition 1.1.13 (d)). Finally, the pre-periodic points for the remaining periodic points are those points in the unit interval with *eventually periodic* continued fraction expansions, that is, numbers of the form $x = [x_1, \ldots, x_m, \overline{x_{m+1}, \ldots, x_{m+k}}]$. These eventually periodic expansions are the subject of the following theorem. Before stating the theorem, recall that a *quadratic surd* is an irrational root of a quadratic equation with integer coefficients. These are also called *algebraic numbers of degree two*.

Theorem 1.2.8 (Lagrange's Theorem). *Every quadratic surd has an eventually periodic continued fraction expansion; conversely, every eventually periodic continued fraction represents a quadratic surd.*

Proof. See, for instance, Theorem 28 in Khintchine [Khi64]. □

From this point on, we are most of the time no longer interested in the rational numbers, as they are simply the countable set of pre-periodic points for the trivial fixed point 0. In other words, from here on all continued fraction expansions are assumed to be infinite. We will denote the irrational points in the unit interval by $\mathbb{I} := [0, 1] \setminus \mathbb{Q}$.

1.2.2 Symbolic dynamics

Recall Proposition 1.2.7, where we showed how the Gauss map acts on the continued fraction expansions. In this section, we will introduce some more general theory which will illuminate the idea behind this proposition, namely, the beginnings of the theory of *symbolic dynamics* and its connections to topological dynamical systems which admit a Markov partition (see Section 1.2.5 for the definition of Markov partitions in the context of interval maps). We will only provide a very short introduction, and for more details we refer, for example, to Lind and Marcus [LM95].

Definition 1.2.9.

(a) Let E be a countable, possibly infinite set containing at least two elements. The set E will be referred to as an *alphabet*. The elements of E will be called *letters* or *symbols*.

(b) For each $n \in \mathbb{N}$ we shall denote by E^n the set of all *words* comprising n letters from the alphabet E. For later convenience, we also denote the *empty word* (that is, the word having no letters) by ε. For instance, if $E = \{0, 1\}$ then

$$E^1 = E \text{ and } E^2 = \{(0,0),(0,1),(1,0),(1,1)\}, \text{ whereas}$$
$$E^3 = \{(0,0,0),(1,0,0),(0,1,0),(0,0,1),(1,1,0),(1,0,1),(0,1,1),(1,1,1)\}.$$

(c) We will denote by $E^* := \bigcup_{n\in\mathbb{N}} E^n$ the set of all finite non-empty words over the alphabet E. The set of all infinite words will be denoted by $E^{\mathbb{N}}$. In other words,

$$E^{\mathbb{N}} := \{\omega = (\omega_i)_{i=1}^{\infty} : \omega_i \in E \text{ for all } i \in \mathbb{N}\}.$$

(d) We define the *length* $|\omega|$ of a word ω to be the number of letters of which it consists. That is, for every $\omega \in E^*$, the length of ω is the unique $n \in \mathbb{N}$ such that $\omega \in E^n$. For $\omega \in E^{\mathbb{N}}$, we have that $|\omega| = \infty$. Furthermore, $|\varepsilon| = 0$.

(e) If $\omega \in E^* \cup E^{\mathbb{N}}$ and $n \in \mathbb{N}$ does not exceed the length of ω, we define the *initial block* $\omega|_n$ to be the initial n-length word of ω, that is, the subword $\omega_1 \omega_2 \ldots \omega_n$.

(f) Given two words $\omega, \tau \in E^* \cup E^{\mathbb{N}}$, we define their *wedge* $\omega \wedge \tau \in \{\varepsilon\} \cup E^* \cup E^{\mathbb{N}}$ to be their longest common initial block. For example, if we again let $E = \{0, 1\}$ and we have words $\omega = (0,0,1,0,1,\ldots)$ and $\tau = (0,0,1,1,0,\ldots)$, then $\omega \wedge \tau = (0,0,1)$. On the other hand, if $\gamma = (1,0,1,0,1,\ldots)$ then $\omega \wedge \gamma = \varepsilon$. Of course, if two (finite or infinite) words ω and τ are equal, then $\omega \wedge \tau = \omega = \tau$.

Let us now introduce a metric on the space $E^{\mathbb{N}}$ which reflects the idea that two words are close if they share a long initial block. In other words, the longer their common initial subword, the closer two words are. We leave the proof that this genuinely defines a metric, in fact an ultrametric, as an exercise (see Exercise 1.6.10).

Definition 1.2.10. Let the metric $d : E^{\mathbb{N}} \times E^{\mathbb{N}} \to [0, 1]$ be defined by $d(\omega, \tau) := 2^{-|\omega \wedge \tau|}$.

If ω and τ have no common initial block, then $\omega \wedge \tau = \varepsilon$. Thus, $|\omega \wedge \tau| = 0$ and $d(\omega, \tau) = 1$. On the other hand, if the two infinite words ω and τ are such that $\omega = \tau$, then $|\omega \wedge \tau| = \infty$ and we define $(1/2)^{+\infty} := 0$.

Definition 1.2.11. Given a finite word $\omega \in E^*$, the *cylinder set* $[\omega]$ generated by ω is the set of all infinite words with initial block ω, that is,

$$[\omega] := \{\tau \in E^{\mathbb{N}} : \tau|_{|\omega|} = \omega\} = \{\tau \in E^{\mathbb{N}} : \tau_i = \omega_i \text{ for all } 1 \le i \le |\omega|\}.$$

We now introduce the shift map, which is defined by dropping the first letter of each word and shifting all the remaining letters one place to the left.

Definition 1.2.12. The *shift map* $\sigma : E^{\mathbb{N}} \to E^{\mathbb{N}}$ is defined by setting $\sigma(w) = \sigma((w_i)_{i\geq 1}) := (w_{i+1})_{i\geq 1}$. That is,

$$\sigma((w_1, w_2, w_3, \ldots)) := (w_2, w_3, w_4, \ldots)$$

The shift map is $\#E$-to-one on $E^{\mathbb{N}}$, where $\#E$ denotes the cardinality of E. In other words, each infinite word has $\#E$ preimages under the shift map. In particular, if E is countably infinite, it follows that σ is countable-to-one. Indeed, given any letter $e \in E$ and any infinite word $w \in E^{\mathbb{N}}$, the concatenation $ew = (e, w_1, w_2, w_3, \ldots)$ of e with w is a preimage of w under the shift map, since $\sigma(ew) = w$.

Note that the shift map is continuous, since two words that are close share a long initial block, and thus their images under the shift map, which result from dropping their first letters, will also share a long initial block. More precisely, whenever $d(w, \tau) < 1$, that is, whenever $|w \wedge \tau| \geq 1$, we have that

$$d(\sigma w, \sigma \tau) = 2^{-|\sigma w \wedge \sigma \tau|} = 2^{-|w \wedge \tau|+1} = 2 \cdot 2^{-|w \wedge \tau|} = 2d(w, \tau).$$

It is evident that if $w = \tau$, then both $d(w, \tau)$ and $d(\sigma(w), \sigma(\tau))$ are equal to zero.

The system $(E^{\mathbb{N}}, \sigma)$ will be referred to as the *full shift system*. Let us now consider another useful construction, that of *sub-shifts*. These are restrictions of the shift map to certain closed and shift-invariant subsets of $E^{\mathbb{N}}$ and are easiest to define in terms of an *incidence matrix*, that is, an $(E \times E)$-matrix consisting entirely of 0s and 1s.

Definition 1.2.13. Let $A = (A_{ij})_{i,j \in E}$ be an incidence matrix. The set of all infinite *A-admissible words* is defined to be

$$E_A^{\mathbb{N}} := \{w \in E^{\mathbb{N}} : A_{w_n w_{n+1}} = 1, \text{ for all } n \in \mathbb{N}\}.$$

Note that if the n-th row of A does not contain any 1, then no word can contain the letter n. This letter then may as well be thrown out of the alphabet. We will therefore impose the condition that every row of A contains at least one 1. Further, notice that if all the entries of the incidence matrix A are equal to 1, then $E_A^{\mathbb{N}} = E^{\mathbb{N}}$. However, if A has at least one 0 entry, then $E_A^{\mathbb{N}}$ is a proper subset of $E^{\mathbb{N}}$, called a sub-shift (of finite type). In particular, if A is the identity matrix then $E_A^{\mathbb{N}} = \{(e, e, e, \ldots) : e \in E\}$, that is, $E_A^{\mathbb{N}}$ is the set of all constant words, which are the fixed points of σ in $E^{\mathbb{N}}$. We will now consider a more interesting sub-shift.

Example 1.2.14. Let $E = \{0, 1\}$ and let

$$A = \begin{bmatrix} 1 & 1 \\ 1 & 0 \end{bmatrix}.$$

Then $E_A^{\mathbb{N}}$ consists of all those points where a 0 can be followed by either a 0 or a 1, but a 1 can only be followed by a 0. This example is known as the *golden mean shift.* One reason for this name can be discovered on considering the cardinality of the sets E_A^n, for $n \in \mathbb{N}$. We have (and the reader is advised to draw a diagram of the A-admissible words to see why) that the sequence $(\# E_A^n)_{n\geq 1}$ coincides with the sequence $(2, 3, 5, 8, 13, 21, 34, \ldots)$. This latter sequence is, of course, the sequence of Fibonacci numbers starting with $f_2 = 2$ and $f_3 = 3$ instead of $f_0 = 1$ and $f_1 = 1$. Observe that we have for the sequence of convergents $(p_n/q_n)_{n\geq 1}$ to $\gamma = (1 + \sqrt{5})/2$ that $p_n/q_n = f_{n+2}/f_{n+1}$ for each $n \in \mathbb{N}$.

In order to define the shift map on a sub-shift space $E_A^{\mathbb{N}}$, we must first verify that these spaces are σ-*invariant*, which means that $\sigma(E_A^{\mathbb{N}}) \subseteq E_A^{\mathbb{N}}$. So, let $\omega \in E_A^{\mathbb{N}}$. Then $A_{\omega_n \omega_{n+1}} = 1$ for every $n \in \mathbb{N}$. In particular, $A_{(\sigma\omega)_n (\sigma\omega)_{n+1}} = A_{\omega_{n+1} \omega_{n+2}} = 1$ for all $n \in \mathbb{N}$. Thus, $\sigma\omega \in E_A^{\mathbb{N}}$ and it therefore follows that $E_A^{\mathbb{N}}$ is σ-invariant. Therefore, the restriction $\sigma : E_A^{\mathbb{N}} \to E_A^{\mathbb{N}}$ is well defined.

1.2.3 A return to the Gauss map

Let us now return to the Gauss map. In light of the above description of symbolic dynamics and the shift map, we can now rephrase Proposition 1.2.7 in the following way: The Gauss map acts on the irrational points of the unit interval like the full shift map acts on the space $\mathbb{N}^{\mathbb{N}}$. More precisely, we have that the topological dynamical systems $(\mathbb{I}, G)$ and $(\mathbb{N}^{\mathbb{N}}, \sigma)$ are topologically conjugate under the map $h : \mathbb{I} \to \mathbb{N}^{\mathbb{N}}$ defined by $h([x_1, x_2, x_3, \ldots]) = (x_1, x_2, x_3, \ldots) \in \mathbb{N}^{\mathbb{N}}$. In other words, we obtain the following commuting diagram:

$$
\begin{array}{ccc}
\mathbb{I} & \xrightarrow{\;G\;} & \mathbb{I} \\
{\scriptstyle h}\downarrow & & \downarrow{\scriptstyle h} \\
\mathbb{N}^{\mathbb{N}} & \xrightarrow[\;\sigma\;]{} & \mathbb{N}^{\mathbb{N}}
\end{array}
$$

We leave it as an exercise for the reader to verify that the map h is genuinely a conjugacy map between the Gauss and shift systems (see Exercise 1.6.12).

Recall the sequence of convergents to each irrational number $x \in [0, 1]$ introduced in Subsection 1.1.1; these are the equivalent of initial blocks for the Gauss map and so we can use them to define Gauss cylinder sets.

Definition 1.2.15. For each choice $x_1, \ldots, x_k \in \mathbb{N}$, define the *k-th level Gauss cylinder set* $C(x_1, \ldots, x_k)$ by

$$
C(x_1, \ldots, x_k) := \{[y_1, y_2, \ldots] : y_i = x_i \text{ for } 1 \leq i \leq k\}.
$$

The first level of these cylinder sets are given by the sets, for $n \in \mathbb{N}$,

$$C(n) = \{[n, x_2, x_3, \ldots] : x_i \in \mathbb{N}, i \geq 2\}.$$

For fixed $n \in \mathbb{N}$, any point $x \in C(n)$ is of the form

$$x = \frac{1}{n + [y_1, y_2, \ldots]},$$

for some $y_1, y_2, \ldots \in \mathbb{N}$. Therefore, since the values of $[y_1, y_2, \ldots]$ lie between 0 and 1, we see that such a point is bigger than $1/(n+1)$ and smaller than $1/n$. Therefore, $C(n) = (1/(n+1), 1/n) \cap \mathbb{I}$. In general, if we fix some $x_1, \ldots, x_k \in \mathbb{N}$ and let $x \in C(x_1, \ldots, x_k)$, we have that

$$x = \cfrac{1}{x_1 + \cfrac{1}{\ddots + \cfrac{1}{x_k + [y_1, y_2, \ldots]}}}$$

and again, since the values of $[y_1, y_2, \ldots]$ range from 0 to 1, we infer that x can take any value in the set $([x_1, \ldots, x_k], [x_1, \ldots, (x_k + 1)])_{\pm} \cap \mathbb{I}$, where the notation $(\cdot, \cdot)_{\pm}$ indicates that the rational number $[x_1, \ldots, x_k]$ may be the left or right endpoint of $C(x_1, \ldots, x_k)$ depending upon whether k is even or odd, respectively. Thus, for the Lebesgue measure $\lambda(C(x_1, \ldots, x_k))$ of this set, we find that

$$\lambda(C(x_1, \ldots, x_k)) = \left| \frac{p_k}{q_k} - \frac{p_k + p_{k-1}}{q_k + q_{k-1}} \right| = \frac{1}{q_k^2 \left(1 + \dfrac{q_{k-1}}{q_k}\right)}.$$

Here, the last equality comes from Theorem 1.1.1 (c). From this, we immediately obtain that

$$\frac{1}{2q_k^2} \leq \lambda(C(x_1, \ldots, x_k)) \leq \frac{1}{q_k^2}. \tag{1.7}$$

It is of interest to calculate the approximate proportion of the k-th level cylinder set $C(x_1, \ldots, x_k)$ that is occupied by each of the $(k + 1)$-th level cylinder sets $C(x_1, \ldots, x_k, n)$. Notice that the endpoints of the interval $C(x_1, \ldots, x_k, n)$ are given by $[x_1, \ldots, x_k, n]$ and $[x_1, \ldots, x_k, n + 1]$ and, with the aid of Theorem 1.1.1, we can write these in terms of the rational numbers $p_{k-1}/q_{k-1} = [x_1, \ldots, x_{k-1}]$ and $p_k/q_k = [x_1, \ldots, x_k]$ as follows:

$$[x_1, \ldots, x_k, n] = \frac{np_k + p_{k-1}}{nq_k + q_{k-1}} \quad \text{and} \quad [x_1, \ldots, x_k, n + 1] = \frac{(n+1)p_k + p_{k-1}}{(n+1)q_k + q_{k-1}}.$$

Utilising Theorem 1.1.1 (c) once more, we obtain that

$$
\begin{aligned}
\lambda(C(x_1,\ldots,x_k,n)) &= \left| \frac{np_k+p_{k-1}}{nq_k+q_{k-1}} - \frac{(n+1)p_k+p_{k-1}}{(n+1)q_k+q_{k-1}} \right| \\
&= \left| \frac{n(p_kq_{k-1}-q_kp_{k-1})+(n+1)(p_{k-1}q_k-p_kq_{k-1})}{(nq_k+q_{k-1})((n+1)q_k+q_{k-1})} \right| \\
&= \frac{1}{n^2q_k^2 \left(1+\dfrac{q_{k-1}}{nq_k}\right)\left(1+\dfrac{1}{n}+\dfrac{q_{k-1}}{nq_k}\right)}
\end{aligned}
$$

Thus, it follows that

$$
\frac{\lambda(C(x_1,\ldots,x_k,n))}{\lambda(C(x_1,\ldots,x_k))} = \frac{1}{n^2} \cdot \frac{1+\dfrac{q_{k-1}}{q_k}}{\left(1+\dfrac{q_{k-1}}{nq_k}\right)\left(1+\dfrac{1}{n}+\dfrac{q_{k-1}}{nq_k}\right)}.
$$

One easily verifies that the second ratio on the right-hand side of the above equality is bounded above by 2 and below by $1/3$, so finally we find that

$$
\frac{1}{3n^2} \le \frac{\lambda(C(x_1,\ldots,x_k,n))}{\lambda(C(x_1,\ldots,x_k))} \le \frac{2}{n^2}. \tag{1.8}
$$

1.2.4 Elementary metrical Diophantine analysis

Let us now consider another area of elementary number theory related to the continued fraction expansion. In this subsection, we will prove some results concerning the Lebesgue measure λ of various sets of numbers, beginning with the set of badly approximable numbers that was introduced in Definition 1.1.15. The proofs will utilise the Gauss cylinder sets defined above.

Theorem 1.2.16. *Where $\mathcal{B}$ denotes the set of badly approximable numbers, we have that*

$$
\lambda(\mathcal{B}) = 0.
$$

Proof. For each N and $n \in \mathbb{N}$, define the sets

$$
A_N^{(n)} := \{[x_1,x_2,\ldots] \in \mathbb{I} : x_i < N \text{ for all } 1 \le i \le n\}.
$$

We aim to show that

$$
\lim_{n\to\infty} \lambda\left(A_N^{(n)}\right) = 0.
$$

To begin, note that $A_N^{(n+1)} \subset A_N^{(n)}$ for each $n \in \mathbb{N}$ and

$$A_N^{(n+1)} = \bigcup_{\substack{(x_1,\ldots,x_n) \\ x_i < N, 1 \le i \le n}} \bigcup_{k:k<N} C(x_1,\ldots,x_n,k).$$

Then,

$$\lambda\left(\bigcup_{k:k<N} C(x_1,\ldots,x_n,k)\right) = \left| \frac{p_n + p_{n-1}}{q_n + q_{n-1}} - \frac{p_n N + p_{n-1}}{q_n N + q_{n-1}} \right|$$

$$= \frac{N-1}{q_n^2(1+s_n)(N+s_n)}$$

$$< \frac{N-1}{q_n^2 N(1+s_n)} = \frac{N-1}{N}\lambda(C(x_1,\ldots,x_n)).$$

Thus,

$$\lambda\left(A_N^{(n+1)}\right) = \lambda\left(\bigcup_{\substack{(x_1,\ldots,x_n) \\ x_i < N, 1 \le i \le n}} \bigcup_{k:k<N} C(x_1,\ldots,x_n,k)\right)$$

$$= \sum_{\substack{(x_1,\ldots,x_n) \\ x_i < N, 1 \le i \le n}} \lambda\left(\bigcup_{k:k<N} C(x_1,\ldots,x_n,k)\right)$$

$$\le \sum_{\substack{(x_1,\ldots,x_n) \\ x_i < N, 1 \le i \le n}} \frac{N-1}{N}\lambda(C(x_1,\ldots,x_n)) = \left(1 - \frac{1}{N}\right)\lambda\left(A_N^{(n)}\right)$$

Applying this estimate n times, we arrive at

$$\lambda\left(A_N^{(n+1)}\right) \le \left(1 - \frac{1}{N}\right)\lambda\left(A_N^{(n)}\right) \le \cdots \le \left(1 - \frac{1}{N}\right)^n \lambda\left(A_N^{(1)}\right).$$

Therefore, $\lim_{n\to\infty} \lambda\left(A_N^{(n)}\right) = 0$. Now, if we define the set A_N by setting

$$A_N := \{[x_1, x_2, \ldots] \in \mathbb{I} : x_i < N \text{ for all } i \in \mathbb{N}\}$$

and notice that $A_N \subset A_N^{(n)}$ for all $n \in \mathbb{N}$, it follows that for all $N \in \mathbb{N}$,

$$\lambda(A_N) = 0.$$

Finally, observing that $B = \bigcup_{N \in \mathbb{N}} A_N$, we have

$$\lambda(B) = \lambda\left(\bigcup_{N \in \mathbb{N}} A_N\right) \le \sum_{N \in \mathbb{N}} \lambda(A_N) = 0.$$

$\square$

Corollary 1.2.17. *Let* $\mathcal{W} := \{[x_1, x_2, \ldots] \in \mathbb{I} : \limsup_{n \to \infty} x_n = \infty\}$. *Then,*

$$\lambda(\mathcal{W}) = 1.$$

Proof. The set $\mathcal{W}$ is simply the complement of the set $\mathcal{B}$ of badly approximable numbers. □

We have now seen that the set of badly approximable numbers does not contribute to sets of irrational numbers of positive Lebesgue measure. Hence, if we want to investigate sets of positive measure, then we have to look for irrationals which are more rapidly approximated by their approximants than is the case for badly approximable irrationals. Our next aim is to prove a theorem, originally due to Borel and Bernstein[1], which will give us some information in this direction. In order to prove this theorem, we will need the following well-known and extremely useful result. The proof is neither long nor complicated, so we include it here for completeness.

Lemma 1.2.18 (Borel–Cantelli Lemma). *Let* $(C_n)_{n \geq 1}$ *be a sequence of Borel-measurable subsets of* $[0, 1]$ *and define the* lim-sup *set* $C_\infty := \limsup_{n \to \infty} C_n$ *to be*

$$\limsup_{n \to \infty} C_n := \bigcap_{n \geq 1} \bigcup_{m \geq n} C_m = \{x \in [0, 1] : x \in C_n \text{ for infinitely many } n \in \mathbb{N}\}.$$

Then, if $\sum_{n=1}^{\infty} \lambda(C_n) < \infty$, *we have that*

$$\lambda(C_\infty) = 0.$$

Proof. The convergence of $\sum_{n=1}^{\infty} \lambda(C_n)$ implies that for each $\varepsilon > 0$, there exists some $n(\varepsilon) \in \mathbb{N}$ such that

$$\sum_{n \geq n(\varepsilon)} \lambda(C_n) < \varepsilon.$$

By the definition of C_∞, we have that

$$C_\infty \subset \bigcup_{n \geq n(\varepsilon)} C_n.$$

Therefore, for arbitrary $\varepsilon > 0$, we obtain that

$$\lambda(C_\infty) \leq \lambda \left(\bigcup_{n \geq n(\varepsilon)} C_n \right) \leq \sum_{n \geq n(\varepsilon)} \lambda(C_n) < \varepsilon.$$

Letting ε tend to zero finishes the proof. □

---

1 Borel's original article [Bor09] contained a mistake, which was observed and corrected by Bernstein [Ber12b, Ber12a]

We are now in a position to prove the second and final main theorem of this subsection.

Theorem 1.2.19 (Borel–Bernstein Theorem). *For a function $\varphi : \mathbb{N} \to (0, \infty)$ we set*

$$\mathcal{W}_\varphi := \{[x_1, x_2, \ldots] \in \mathbb{I} : x_n > \varphi(n) \text{ for infinitely many } n \in \mathbb{N}\}.$$

(a) *If the series $\sum_{n=1}^{\infty} 1/\varphi(n)$ diverges, then*

$$\lambda(\mathcal{W}_\varphi) = 1,$$

(b) *If the series $\sum_{n=1}^{\infty} 1/\varphi(n)$ converges, then*

$$\lambda(\mathcal{W}_\varphi) = 0.$$

Proof. To prove part (a) we show that the complement of $\mathcal{W}_\varphi$ has measure zero. The proof follows along the same lines as the proof of Theorem 1.2.16. As before, we obtain that

$$\lambda \left(\bigcup_{1 \leq k < \varphi(n+1)} C(x_1, \ldots, x_n, k) \right) < \left(1 - \frac{1}{\varphi(n+1)} \right) \lambda \left(C(x_1, \ldots, x_n) \right).$$

Therefore, on setting $\mathcal{B}_\varphi^{(m,n)} := \{[x_1, x_2, \ldots] \in \mathbb{I} : x_i \leq \varphi(i) \text{ for all } m \leq i \leq n\}$, we obtain that

$$\lambda \left(\mathcal{B}_\varphi^{(m,n+1)} \right) < \left(1 - \frac{1}{\varphi(n+1)} \right) \lambda \left(\mathcal{B}_\varphi^{(m,n)} \right) < \cdots$$
$$< \prod_{k=m}^{n} \left(1 - \frac{1}{\varphi(k+1)} \right) \lambda \left(\mathcal{B}_\varphi^{(m,m)} \right).$$

Using the fact that $1 - x < e^{-x}$ for each $0 < x < 1$, we have that

$$\lambda \left(\mathcal{B}_\varphi^{(m,n+1)} \right) < e^{-\sum_{k=m}^{n} 1/\varphi(k+1)} \lambda \left(\mathcal{B}_\varphi^{(m,m)} \right),$$

which implies, since by assumption the series $\sum_{k=m}^{n} 1/\varphi(k+1)$ gets arbitrarily large, that for each $m \in \mathbb{N}$,

$$\lim_{n \to \infty} \lambda \left(\mathcal{B}_\varphi^{(m,n)} \right) = 0.$$

Hence, as $\mathcal{B}_\varphi^m := \{[x_1, x_2, \ldots] \in \mathbb{I} : x_i \leq \varphi(i) \text{ for all } i \geq m\} \subset \mathcal{B}_\varphi^{(m,n)}$ for all $n \in \mathbb{N}$, we finally obtain that $\lambda(\mathcal{B}_\varphi^m) = 0$. Consequently,

$$\lambda(\mathcal{W}_\varphi^C) = \lambda \left(\bigcup_{m \geq 1} \mathcal{B}_\varphi^m \right) = 0.$$

It remains to prove part (b). For this, we aim to use the Borel–Cantelli Lemma. To that end, define the sets $\mathcal{W}_\varphi^{(n)}$, for $n \in \mathbb{N}$, by setting

$$\mathcal{W}_\varphi^{(n)} := \{[x_1, x_2, \ldots] \in \mathbb{I} : x_n > \varphi(n)\}.$$

It is then clear that

$$\mathcal{W}_\varphi = \{x \in \mathbb{I} : x \in \mathcal{W}_\varphi^{(n)} \text{ for infinitely many } n \in \mathbb{N}\}.$$

So, in order to apply the Borel–Cantelli Lemma, it suffices to show that the series $\sum_{n=1}^\infty \lambda\left(\mathcal{W}_\varphi^{(n)}\right)$ converges. Indeed,

$$\lambda\left(\mathcal{W}_\varphi^{(n+1)}\right) = \lambda\left(\bigcup_{(x_1,\ldots,x_n)\in\mathbb{N}^n} \bigcup_{k>\varphi(n+1)} C(x_1,\ldots,x_n,k)\right) \tag{1.9}$$

and

$$\lambda\left(\bigcup_{k>\varphi(n+1)} C(x_1,\ldots,x_n,k)\right) = \left|\frac{p_n\varphi(n+1)+p_{n-1}}{q_n\varphi(n+1)+q_{n-1}} - \frac{p_n}{q_n}\right|$$

$$= \frac{1}{q_n^2(1+s_n)}\frac{1+s_n}{\varphi(n+1)+s_n}$$

$$< \frac{2}{\varphi(n+1)}\lambda(C(x_1,\ldots,x_n)). \tag{1.10}$$

Thus, on combining (1.9) and (1.10), we have that

$$\lambda\left(\mathcal{W}_\varphi^{(n+1)}\right) < \frac{2}{\varphi(n+1)}\sum_{(x_1,\ldots,x_n)\in\mathbb{N}^n}\lambda(C(x_1,\ldots,x_n)) = \frac{2}{\varphi(n+1)}$$

and so,

$$\sum_{n=1}^\infty \lambda\left(\mathcal{W}_\varphi^{(n+1)}\right) < 2\sum_{n=1}^\infty 1/\varphi(n) < \infty.$$

This finishes the proof. $\qquad\square$

Corollary 1.2.20. *For λ-a.e. $[x_1, x_2, x_3, \ldots] \in [0, 1]$, we have that*

$$x_n > n\log n, \quad \text{for infinitely many } n \in \mathbb{N},$$

whereas for every $\varepsilon > 0$, we have that

$$x_n < n(\log n)^{1+\varepsilon}, \quad \text{for all sufficiently large } n \in \mathbb{N}.$$

Proof. This follows from Theorem 1.2.19 immediately on the observation that

$$\sum_{n=1}^{\infty} \frac{1}{n \log n} = \infty \text{ and } \sum_{n=1}^{\infty} \frac{1}{n(\log n)^{1+\varepsilon}} < \infty, \text{ for all } \varepsilon > 0.$$

$\square$

1.2.5 Markov partitions for interval maps

In this subsection, we introduce the idea of a Markov partition. This notion will be used repeatedly to build symbolic codings for our various examples. We will restrict the discussion to maps defined on the closed unit interval $[0, 1]$ which are either already continuous or become continuous when considered as functions on the circle $\mathbb{R}/\mathbb{Z} \simeq [0, 1)$ identifying the points 0 and 1, with the exception of a countable set of points (denoted by $\mathcal{E}$) where the map can be discontinuous. We will refer to such maps as *Markov interval maps*. To make this clearer, all the possibilities are covered by considering the tent map, the doubling map (modulo 1), as in Example 1.2.4 (see also Fig. 1.2), and the Gauss map (modulo 1) (see Fig. 1.1), for which the exceptional set is $\mathcal{E} = \{0\}$.

Definition 1.2.21. Let $T : J \to J$ be a map as described above, that is $J := [0, 1]$ or $\mathbb{R}/\mathbb{Z}$. Further, let $\mathbb{M} := \{A_i : i \in E \subset \mathbb{N}\}$ be a countable collection of open non-empty sub-intervals of J and define $\mathcal{E} := J \setminus \bigcup_{i \in E} \overline{A}_i$, where we suppose that the set $\mathcal{E}$ of exceptional points is at most countable. In here, $\overline{A}$ denotes the topological closure

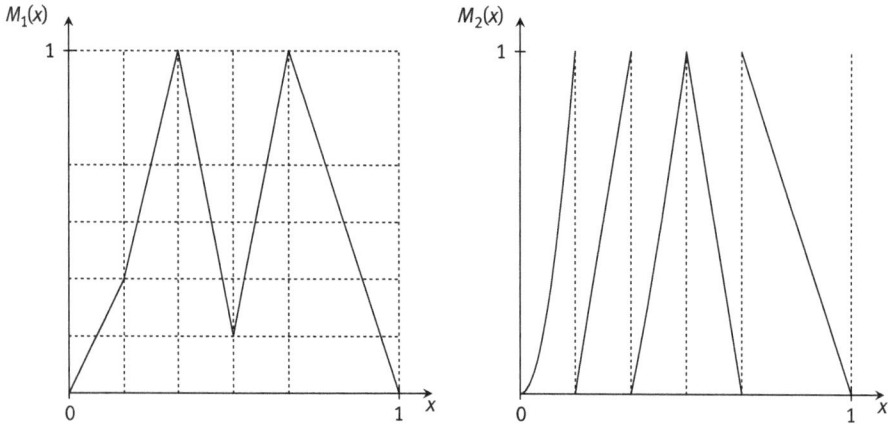

Fig. 1.2. The graph of two Markov interval maps. The map M_1 is continuous on $[0, 1]$ whereas M_2 has to be considered as a continuous function on the circle $\mathbb{R}/\mathbb{Z} \simeq [0, 1)$.

of the subset A in J. Then, the collection $\mathbb{M}$ is said to be a *Markov partition* for T if it satisfies the following properties:

(a) $T|_{\overline{A_i}} : \overline{A_i} \to T(\overline{A_i})$ defines a homeomorphism for each $i \in E$.

(b) $A_i \cap A_j = \varnothing$ for all $i, j \in E$ with $i \neq j$.

(c) If $T(A_i) \cap A_j \neq \varnothing$ for some $i, j \in E$, then $A_j \subset T(A_i)$.

Example 1.2.22. We give here the most natural Markov partition for the Gauss map, $G : \mathbb{R}/\mathbb{Z} \to \mathbb{R}/\mathbb{Z}$. For each $i \in \mathbb{N}$, let $A_i := (1/(i+1), 1/i)$, so that $\mathbb{M}$ is the collection of sets $\mathbb{M} = \{(1/(i+1), 1/i) : i \in \mathbb{N}\}$. Then the set $\mathcal{E}$ is equal to the singleton $\{0\}$. Notice that $\mathbb{M}$ in fact coincides with the family of first level Gauss cylinder sets, up to sets of measure zero. We must check that the three properties defining a Markov partition are satisfied for this choice of $\mathbb{M}$. Property (a) is clearly satisfied, by the definition of the Gauss map. Property (b) is also straightforward to check. For property (c), it is enough to notice that for all $i \in \mathbb{N}$ we have

$$G(A_i) = (0, 1).$$

Let us now consider collections $\mathcal{U}$ of subsets of X and define some operations on these collections.

Definition 1.2.23. Let $T : X \to X$ be some given map.

(a) The *join* $\mathcal{U} \vee \mathcal{V}$ of two collections $\mathcal{U}$ and $\mathcal{V}$ of X is defined to be

$$\mathcal{U} \vee \mathcal{V} := \{U \cap V : U \in \mathcal{U}, V \in \mathcal{V}\}.$$

This definition is extended to finitely many collections in the natural way.

(b) We define the *preimage* of a collection $\mathcal{U}$ under the map T to be the collection consisting of all the preimages of the elements of $\mathcal{U}$ under T, that is,

$$T^{-1}\mathcal{U} := \{T^{-1}(U) : U \in \mathcal{U}\}.$$

The collection $T^{-n}(\mathcal{U})$ is defined inductively to be the preimage of the collection $T^{-(n-1)}(\mathcal{U})$, in other words, $T^{-n}(\mathcal{U}) := T^{-1}(T^{-(n-1)}(\mathcal{U}))$.

(c) For each $n \in \mathbb{N}$, we define the *n-th refinement* $\mathcal{U}^n$ of a collection $\mathcal{U}$ to be

$$\mathcal{U}^n := \bigvee_{k=0}^{n-1} T^{-k}(\mathcal{U}) = \mathcal{U} \vee T^{-1}(\mathcal{U}) \vee \cdots \vee T^{-(n-1)}(\mathcal{U}).$$

These operations on collections of sets can now be used to define a certain type of Markov partition.

Definition 1.2.24. A Markov partition $\mathbb{M}$ for a dynamical system $([0, 1], T)$ is said to be *shrinking* if the diameter of the largest element in each refinement $\mathbb{M}^n$ shrinks to zero as n tends to infinity. That is, $\mathbb{M}$ is shrinking provided that

$$\limsup_{n\to\infty} \{\operatorname{diam}(M) : M \in \mathbb{M}^n\} = 0.$$

Each shrinking Markov partition gives rise to a canonical coding, in the following way. (Here we are following [Adl98], and we refer to this paper for further discussion.)

Definition 1.2.25. Let $\mathbb{M} := \{A_i : i \in E \subseteq \mathbb{N}\}$ be a shrinking Markov partition for a dynamical system $([0, 1], T)$ and let $A = (A_{ij})$ be the $E \times E$ incidence matrix defined, for each $i, j \in E$, by $A_{ij} = 1$ if and only if $T(A_i) \cap A_j \neq \varnothing$. Then, where $E_A^{\mathbb{N}}$ is the sub-shift consisting of all A-admissible words, the *coding* associated with $\mathbb{M}$ is given by the map

$$\pi : E_A^{\mathbb{N}} \to [0, 1] \setminus \bigcup_{n=0}^{\infty} T^{-n}(\mathcal{E})$$

defined by

$$\pi((\omega_1, \omega_2, \omega_3, \ldots)) := \bigcap_{n=0}^{\infty} \overline{A_{\omega_1} \cap T^{-1} A_{\omega_2} \cap \cdots \cap T^{-n} A_{\omega_{n+1}}}.$$

Note that this map is well defined, since the intersection on the right-hand side above is a singleton, due to Cantor's Intersection Theorem (since the sequence $\overline{(A_{\omega_1} \cap T^{-1} A_{\omega_2} \cap \cdots \cap T^{-n} A_{\omega_{n+1}})}_{n\geq 0}$ is a nested sequence of compact sets with diameters shrinking to zero; see Exercise 1.6.14 if you have not encountered this useful result before). Due to the restriction that the exceptional set $\mathcal{E}$ be countable, we obtain a unique coding for all but countably many points.

Theorem 1.2.26. *The map π is a factor map from the system $(E_A^{\mathbb{N}}, \sigma)$ to the system $([0, 1] \setminus \bigcup_{n=0}^{\infty} T^{-n}(\mathcal{E}), T)$.*

Proof. Recall that the definition of a factor map is that π should be a continuous surjection from E_A^{∞} to $[0, 1] \setminus \bigcup_{n=0}^{\infty} T^{-n}(\mathcal{E})$ such that $\pi \circ \sigma = T \circ \pi$. It is easily demonstrated that π is uniformly continuous. Indeed, for this it is enough to notice that $\operatorname{diam}(\pi([x_1, \ldots, x_n])) \leq \sup \{\operatorname{diam}(M) : M \in \mathbb{M}^n\} \to 0$, for $n \to \infty$, and uniformly for all admissible n-cylinders.

To see that π is surjective, we argue inductively. Let $x \in [0, 1] \setminus \bigcup_{n=0}^{\infty} T^{-n}(\mathcal{E})$ be an element of $\overline{A_{\omega_1} \cap T^{-1} A_{\omega_2} \cap \cdots \cap T^{-(n-1)} A_{\omega_n}}$ for some $n \in \mathbb{N}$. Since the collection $\{\overline{M} : M \in \mathbb{M}^{n+1}\}$ covers the set

$$\overline{A_{\omega_1} \cap T^{-1} A_{\omega_2} \cap \cdots \cap T^{-(n-1)} A_{\omega_n}} \setminus \bigcup_{n=0}^{\infty} T^{-n}(\mathcal{E}),$$

there must exist (at least) one element $\overline{A_{\omega_1} \cap T^{-1} A_{\omega_2} \cap \cdots \cap T^{-n} A_{\omega_{n+1}}}$ containing x with $T(A_{\omega_n}) \cap A_{\omega_{n+1}} \neq \varnothing$. In this way we construct a point $(\omega_1, \omega_2, \omega_3, \ldots)$ lying in $E_A^{\mathbb{N}}$ and one quickly verifies that $\pi((\omega_1, \omega_2, \omega_3, \ldots))$ is equal to x.

Finally, it remains to show that $\pi \circ \sigma = T \circ \pi$. So, let $\omega = (\omega_1, \omega_2, \omega_3, \ldots) \in E_A^{\mathbb{N}}$ and observe that

$$\pi \circ \sigma(\omega) = \pi((\omega_2, \omega_3, \omega_4, \ldots)) = \bigcap_{n=0}^{\infty} \overline{A_{\omega_2} \cap T^{-1} A_{\omega_3} \cap \cdots \cap T^{-(n-1)} A_{\omega_{n+1}}}.$$

On the other hand, using the continuity of the restriction of T to $\overline{A_{\omega_1}}$ we have that

$$T \circ \pi(\omega) = T \left(\bigcap_{n=0}^{\infty} \overline{A_{\omega_1} \cap T^{-1} A_{\omega_2} \cap \cdots \cap T^{-n} A_{\omega_{n+1}}} \right)$$

$$= \bigcap_{n=0}^{\infty} \overline{T A_{\omega_1} \cap A_{\omega_2} \cap \cdots \cap T^{-(n-1)} A_{\omega_{n+1}}}$$

$$= \bigcap_{n=0}^{\infty} \overline{A_{\omega_2} \cap T^{-1} A_{\omega_3} \cap \cdots \cap T^{-(n-1)} A_{\omega_{n+1}}}$$

where the final equality comes from the fact that $T(A_{\omega_1}) \supset A_{\omega_2}$, since ω belongs to the set $E_A^{\mathbb{N}}$. This finishes the proof. □

Example 1.2.27. Let us return once more to the Gauss map and to its canonical Markov partition given in Example 1.2.22. We can easily see that the Markov partition $\mathbb{M}$ given in that example is shrinking. Indeed, the refinement $\mathbb{M}^n$ coincides with the collection of n-th level Gauss cylinder sets, up to sets of measure zero, and the largest element of this collection is $C(1, \ldots, 1)$. We saw in (1.7) that

$$\frac{1}{2q_n^2} \leq \lambda(C(x_1, \ldots, x_n)) \leq \frac{1}{q_n^2},$$

and since for $C(1, \ldots, 1)$ we have that q_n is equal to the $(n+1)$-th Fibonacci number, it is clear that as n tends to infinity, the diameter of the cylinder set with code consisting of n 1s shrinks to zero.

Since $\mathbb{M}$ is shrinking, we can therefore obtain a coding from it. Recall that the set $\mathcal{E}$ in this case is equal to the singleton $\{0\}$. Therefore, the set of points where the coding is undefined is here simply equal to the set of pre-periodic points of 0, that is, the rational numbers in $[0, 1]$. So, for any point $\omega = (x_1, x_2, x_3, \ldots) \in \mathbb{N}^{\mathbb{N}}$, we have that

$$\{\pi(\omega)\} = A_{x_1} \cap G^{-1}(A_{x_2}) \cap G^{-2}(A_{x_3}) \cap \cdots.$$

In other words, we have that $\pi(\omega)$ lies in $A_{x_1} = (1/(x_1 + 1), 1/x_1)$, $G(\pi(\omega))$ lies in $A_{x_2} = (1/(x_2 + 1), 1/x_2)$, $G^2(\pi(\omega))$ lies in $A_{x_3} = (1/(x_3 + 1), 1/x_3)$ and so on. Thus, the first continued fraction element of $\pi(\omega)$ is equal to x_1, the second is equal to x_2, the third is equal to x_3 and so on. Therefore, we have shown that the coding that arises from the Markov partition $\mathbb{M}$ coincides with the continued fraction expansion of each irrational number.

Note that in this case the map $\pi : \mathbb{N}^{\mathbb{N}} \to \mathbb{I}$ is actually a topological conjugacy map, since the endpoints of all the intervals in $\mathbb{M}$ lie in the set $G^{-1}(0)$ and hence, the map is injective as well as surjective.

Remark 1.2.28. The ideas of this section can extended much further, to the setting of graph-directed Markov systems. For this, we refer the reader to the textbook by Mauldin and Urbanski [MU03].

1.3 The Farey map: definition and topological properties

In this section, we introduce the second of our main examples, namely, the Farey map. First we give the definition and show how the Farey map is related to the Gauss map. We then define a Markov partition for the Farey map and describe the coding associated with it. Finally, in the second subsection, we give some basic topological properties of the Farey map.

1.3.1 The Farey map

Let us now define a transformation on the unit interval that is closely related to the Gauss map.

Definition 1.3.1. The *Farey map* $F : [0, 1] \to [0, 1]$ is defined by

$$F(x) := \begin{cases} x/(1-x) & \text{for } 0 \le x \le \frac{1}{2}; \\ (1-x)/x & \text{for } \frac{1}{2} < x \le 1. \end{cases}$$

The graph of the Farey map is shown in Fig. 1.3.

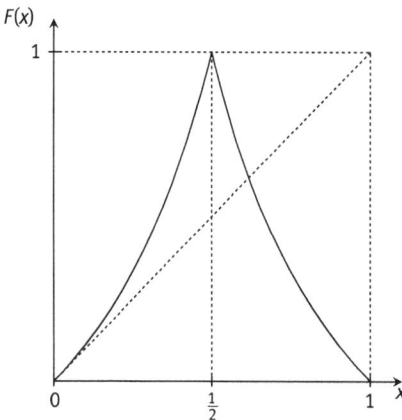

Fig. 1.3. The Farey map, $F : [0, 1] \to [0, 1]$. In the fixed point at 0 the map F has slope 1. The second fixed point lies in $\gamma^* = (\sqrt{5} - 1)/2$.

For later use, let us also give the inverse branches of F. These are the functions $F_0 : (0, 1) \to (0, 1/2)$ and $F_1 : (0, 1) \to (1/2, 1)$ which are defined by

$$F_0(x) := \frac{x}{1+x} \quad \text{and} \quad F_1(x) := \frac{1}{1+x}.$$

It is easily shown that the action of the map F on a point $x = [x_1, x_2, x_3, \ldots] \in (0, 1)$ is as follows:

$$F(x) = \begin{cases} [x_1 - 1, x_2, x_3, \ldots] & \text{if } x_1 > 1; \\ [x_2, x_3, x_4, \ldots] & \text{if } x_1 = 1. \end{cases}$$

For this reason, the Farey map is sometimes referred to as the *slow continued fraction map*, whereas the Gauss map is referred to as the *fast continued fraction map*. We can describe the relationship between the Farey and Gauss maps more precisely. To do this, we introduce the idea of a jump transformation, which is also often referred to as *Schweiger's jump transformation* [Sch95].

Definition 1.3.2. Let the map $\rho : (0, 1] \to \mathbb{N} \cup \{0\}$ be defined by

$$\rho(x) := \inf\{n \geq 0 : F^n(x) \in (1/2, 1]\}.$$

Note that $\rho(x)$ is finite for all $x \in (0, 1]$. Then, let the map $F^* : [0, 1] \to [0, 1]$ be defined by

$$F^*(x) := \begin{cases} F^{\rho(x)+1}(x) & \text{if } x \neq 0; \\ 0 & \text{if } x = 0. \end{cases}$$

The map F^* is said to be the *jump transformation on* $(1/2, 1]$ of F.

Lemma 1.3.3. *The jump transformation F^* of the Farey map coincides with the Gauss map.*

Proof. Fix $n \geq 2$ and let $x = [n, x_2, x_3, \ldots]$, so $x \in (1/(n+1), 1/n]$. Then, we have that

$$F^*(x) = F^n([n, x_2, x_3, \ldots]) = F^{n-1}([n-1, x_2, x_3, \ldots]) = \cdots$$
$$= F([1, x_2, x_3, \ldots]) = [x_2, x_3, \ldots] = G(x).$$

On the other hand, if $x = [1, x_2, x_3, \ldots] \in (1/2, 1]$, then $\rho(x)$ is equal to zero and so we have that $F^*(x) = F(x)$, which is again equal to $G(x)$, since $G|_{(1/2,1]} = F|_{(1/2,1]}$. $\square$

Our next aim is to describe a coding generated by the Farey map. The two open sets $\{B_0 := (0, 1/2), B_1 := (1/2, 1)\}$ form a Markov partition for F. In this case, the exceptional set $\mathcal{E}$ is the empty set. That the three conditions for a Markov partition are satisfied is easy to check, so we leave this as an exercise.

The above Markov partition for the Farey map yields a coding of the numbers in $[0, 1]$ as follows. Each element $\omega = (x_1, x_2, x_3, \ldots) \in \{0, 1\}^{\mathbb{N}}$ is mapped by π to the point $x \in [0, 1]$ given by

$$\{x\} := \bigcap_{n=0}^{\infty} \overline{B_{x_1} \cap F^{-1}B_{x_2} \cap \cdots \cap F^{-n}B_{x_{n+1}}}.$$

We will use the notation $\pi(\omega) = x =: \langle x_1, x_2, x_3, \ldots \rangle \in [0, 1]$. This coding will be referred to as the *Farey coding*. The Farey coding is related to the continued fraction expansion of x in a straightforward way. Indeed, if $x = [x_1, x_2, x_3, \ldots]$, then the Farey coding of x is given by $x = \langle 0^{x_1-1}, 1, 0^{x_2-1}, 1, 0^{x_3-1}, 1, \ldots \rangle$, where 0^n denotes the sequence of $n \in \mathbb{N}$ consecutive appearances of the symbol 0 and 0^0 is understood to mean the appearance of no zeros between two consecutive 1s. This follows directly from the fact that the Gauss map is the jump transformation on $(1/2, 1]$ of the Farey map.

We have described the coding for irrational numbers; let us now consider the rational numbers. We can still define a code for these points, it will just no longer be unique. So, let $x = [x_1, \ldots, x_k]$. In this case we obtain two infinite codings, namely, we have that

$$x = \langle 0^{x_1-1}, 1, 0^{x_2-1}, 1, \ldots, 0^{x_k-1}, 1, 0, 0, 0, \ldots \rangle$$

and

$$x = \langle 0^{x_1-1}, 1, 0^{x_2-1}, 1, \ldots, 0^{x_k-2}, 1, 1, 0, 0, 0, \ldots \rangle.$$

Indeed, take for example the point $1/2$. Observe that $1/2 \in \overline{B_1}$ but also $1/2 \in \overline{B_0}$, so the first entry in the Farey coding for $1/2$ can be either 1 or 0. Then $F(1/2) = 1$ and $F^n(1/2) = 0$ for all $n \geq 2$, therefore we obtain that $1/2 = \langle 1, 1, 0, 0, 0, \ldots \rangle$ and $1/2 = \langle 0, 1, 0, 0, 0, \ldots \rangle$. The coding for any rational number can be deduced from this example, since every rational number is eventually mapped to $1/2$ by the map F (or, to put it another way, the set of rational numbers coincides with the iteration of the point $1/2$ under the inverse branches of the Farey map).

Recall that F acts on $x = [x_1, x_2, x_3, \ldots]$ in the following way:

$$F(x) := \begin{cases} [x_1 - 1, x_2, x_3, \ldots] & \text{for } x_1 \geq 2; \\ [x_2, x_3, x_4 \ldots] & \text{for } x_1 = 1. \end{cases}$$

In particular, this means that if we instead write x in its Farey coding, so $x = \langle y_1, y_2, \ldots \rangle$, then

$$F(x) = \langle y_2, y_3, \ldots \rangle.$$

So, again, the system $(\mathbb{I}, F)$ can be thought of as acting like the shift map σ on the shift space $E^{\mathbb{N}}$, where this time the alphabet $E = \{0, 1\}$ is finite. This is a potential advantage of the map F over the map G in certain situations, as the space $\{0, 1\}^{\mathbb{N}}$ is a compact metric space, whereas $\mathbb{N}^{\mathbb{N}}$ is not.

We are now in a position to define the cylinder sets with respect to the Farey map.

Definition 1.3.4. For each n-tuple $(x_1, \ldots, x_n) \in \{0, 1\}^n$, define the *Farey cylinder set* $\widehat{C}(x_1, \ldots, x_n)$ by setting

$$\widehat{C}(x_1, \ldots, x_n) := \{\langle y_1, y_2, \ldots \rangle : y_k = x_k, \text{ for all } 1 \le k \le n\}.$$

Note that the Farey cylinder sets coincide with the refinements of the Markov partition $\{B_0, B_1\}$ under the inverse branches F_0 and F_1 of F. That is, the cylinder set $\widehat{C}(x_1, \ldots, x_n)$ coincides with the set $\overline{F_{x_1} \circ F_{x_2} \circ \cdots \circ F_{x_n}((0, 1))}$. We will refer to these successive refinements as the *Farey decomposition*.

There is another way to describe these cylinder sets, in terms of the classical construction of Stern–Brocot intervals (cf. [Ste58], [Bro61]). For each $n \ge 0$, the elements of the n-th member of the Stern–Brocot sequence

$$\mathcal{B}_n := \left\{ \frac{s_{n,k}}{t_{n,k}} : k = 1, \ldots, 2^n + 1 \right\}$$

are defined recursively as follows:
- $s_{0,1} := 0$ and $s_{0,2} := t_{0,1} := t_{0,2} := 1$;
- $s_{n+1,2k-1} := s_{n,k}$ and $t_{n+1,2k-1} := t_{n,k}$, for $k = 1, \ldots, 2^n + 1$;
- $s_{n+1,2k} := s_{n,k} + s_{n,k+1}$ and $t_{n+1,2k} := t_{n,k} + t_{n,k+1}$, for $k = 1, \ldots 2^n$.

From this arises the set $\mathcal{T}_n$ of *Stern–Brocot intervals of order* n which is given by

$$\mathcal{T}_n := \left\{ \left[\frac{s_{n,k}}{t_{n,k}}, \frac{s_{n,k+1}}{t_{n,k+1}} \right] : k = 1, \ldots, 2^n \right\}.$$

It is straightforward to check that these intervals are precisely the same intervals as the set

$$\{\widehat{C}(x_1, \ldots, x_n) : x_i \in \{0, 1\}, 1 \le i \le n\}.$$

Returning to the Stern–Brocot sequence, the n-th member of this sequence consists of $2^n + 1$ proper fractions and the n-th member of the sequence can be obtained from the $(n-1)$-th member by adding in the *mediant* of each neighbouring pair, where we remind the reader that the mediant of two rational numbers a/b and a'/b' is by definition the rational number $(a + a')/(b + b')$. The first few of these sequences are given by:

$$\mathcal{B}_0 = \left\{ \frac{0}{1}, \frac{1}{1} \right\}, \quad \mathcal{B}_1 = \left\{ \frac{0}{1}, \frac{1}{2}, \frac{1}{1} \right\}, \quad \mathcal{B}_2 = \left\{ \frac{0}{1}, \frac{1}{3}, \frac{1}{2}, \frac{2}{3}, \frac{1}{1} \right\},$$

$$\mathcal{B}_3 = \left\{ \frac{0}{1}, \frac{1}{4}, \frac{1}{3}, \frac{2}{5}, \frac{1}{2}, \frac{3}{5}, \frac{2}{3}, \frac{3}{4}, \frac{1}{1} \right\},$$

$$\mathcal{B}_4 = \left\{ \frac{0}{1}, \frac{1}{5}, \frac{1}{4}, \frac{2}{7}, \frac{1}{3}, \frac{3}{8}, \frac{2}{5}, \frac{3}{7}, \frac{1}{2}, \frac{4}{7}, \frac{3}{5}, \frac{5}{8}, \frac{2}{3}, \frac{5}{7}, \frac{3}{4}, \frac{4}{5}, \frac{1}{1} \right\}, \ldots$$

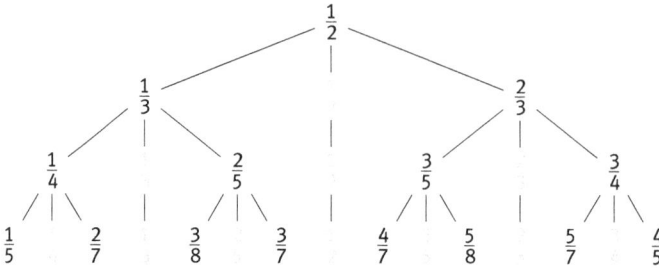

Fig. 1.4. The dyadic *Stern–Brocot tree* with root $s_{1,2}/t_{1,2} = 1/2$. For each $s_{n,2k}/t_{n,2k}$, $n \in \mathbb{N}$ and $k = 1, \ldots, 2^{n-1}$ we have the two offspring $s_{n+1,2k}/t_{n+1,2k}$ and $s_{n+1,2k+2}/t_{n+1,2k+2}$. For each n, the missing elements from $\mathcal{B}_n \backslash \{0, 1\}$ are marked in grey.

This sequence also gives rise to the *Stern–Brocot Tree* as shown in Fig. 1.4. The vertices of the n-th generation of the Stern–Brocot Tree will be denoted by $\mathcal{S}_n := \mathcal{B}_n \backslash \mathcal{B}_{n-1}$, for $n \in \mathbb{N}$ and the sequence $(\mathcal{S}_n)_{n \in \mathbb{N}}$ is called the *even Stern–Brocot sequence*.

Remark 1.3.5. The Markov partition $\{B_0, B_1\}$ given above is not the only reasonable choice for the Farey map. We might instead choose to use the partition $\mathbb{M}$, that is, the partition defined in Example 1.2.22 for the Gauss map. It is not hard to verify that this is also a shrinking Markov partition for the Farey map, and so yields a different coding for the points of $[0, 1]$. In this case, we obtain a coding map $\pi : \mathbb{N}_A^\mathbb{N} \to [0, 1] \backslash \bigcup_{n \in \mathbb{N}} F^{-n}(0)$, where the sub-shift $\mathbb{N}_A^\mathbb{N}$ is determined by the infinite transition matrix given by

$$A_{ij} = \begin{cases} 1 & \text{if } i = 1 \text{ and } j \in \mathbb{N}; \\ 1 & \text{if } i = n \text{ and } j = n - 1, \text{ for } n \geq 2; \\ 0 & \text{otherwise.} \end{cases}$$

This sub-shift is known as the *(infinite) renewal shift*. We will return to the subject of renewal theory in Chapter 3.

1.3.2 Topological properties of the Farey map

We now come to the study of the Farey map from a topological point of view. Let us first introduce another well-studied dynamical system, which we shall shortly show to be topologically conjugate to the Farey map.

Definition 1.3.6. Define the map $T : [0, 1] \to [0, 1]$ by setting

$$T(x) := \begin{cases} 2x & \text{for } x \in [0, 1/2); \\ 2 - 2x & \text{for } x \in [1/2, 1]. \end{cases}$$

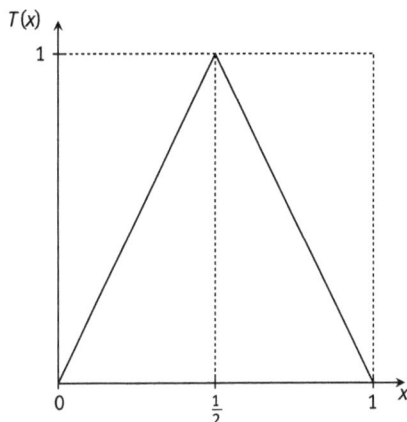

Fig. 1.5. The tent map, $T : [0,1] \to [0,1]$.

The map T is referred to as the *tent map*; the reason for this name is clear on inspection of the graph of T, shown in Fig. 1.5.

It turns out, and we shall prove this shortly, that the conjugating homeomorphism between the Farey map and the tent map coincides with *Minkowski's question-mark function*, which we shall denote by $Q : [0,1] \to [0,1]$. This remarkable function was originally introduced by Minkowski [Min10] and later investigated by Denjoy [Den38] and Salem [Sal43], amongst others. Minkowski's original motivation behind the definition of the function that now bears his name was to highlight the intriguing property of continued fractions that was described in Lagrange's Theorem (see Theorem 1.2.8). Recall that this theorem states that the set of irrational algebraic numbers of degree two corresponds precisely to the set of real numbers that admit an eventually periodic continued fraction expansion. In other words, if $x \in [0, 1]$ can be written as a continued fraction of the form $[x_1, \ldots, x_m, \overline{x_{m+1}, \ldots, x_{m+k}}]$, then x is an irrational root of some quadratic polynomial and, moreover, the converse statement also holds. Minkowski designed the function Q to map the quadratic surds into the non-dyadic rationals in a continuous and order-preserving way (we leave the proof of this to Exercise 1.6.15). The question-mark function is constructed in the following way. First, define $Q(0) = Q(0/1) := 0$ and $Q(1) = Q(1/1) := 1$. Then, define

$$Q\left(\frac{p+p'}{q+q'}\right) := \frac{Q(p/q) + Q(p'/q')}{2}.$$

In other words, the function Q is successively defined on all the rational numbers in the unit interval by taking mediants of those that have already been defined. The definition of Q is extended to all of $[0, 1]$ by continuity (since any uniformly continuous

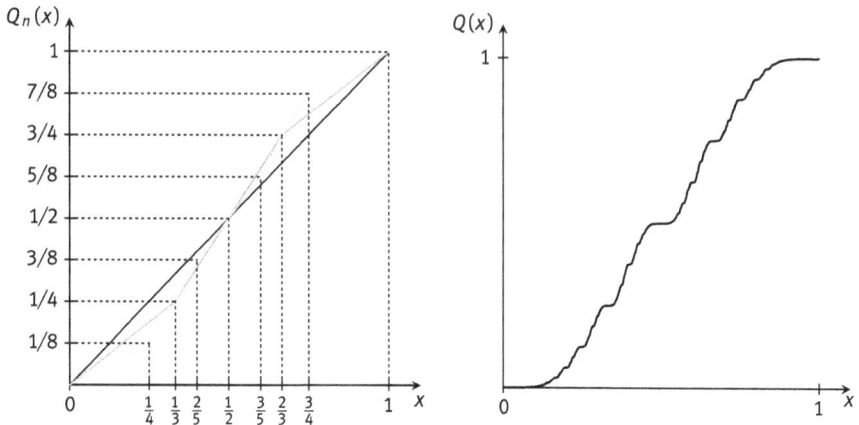

Fig. 1.6. On the left, the graphs of the functions Q_n, $n = 1, 2, 3$, and on the right, an approximation to the graph of the Minkowski question-mark function, $Q : [0, 1] \to [0, 1]$.

function from a dense set of a metric space E into another metric space can be uniquely extended to a continuous function on all of E).

Another way to think about the question-mark function is as a uniform limit of the sequence of piecewise linear functions $(Q_n)_{n \in \mathbb{N}}$, where each $Q_n : [0, 1] \to [0, 1]$ is defined by mapping the n-th level Stern–Brocot fractions, arranged in increasing order, onto the set $\{p/2^n : 0 \le p \le 2^n\}$ and then joining these image points by straight line segments. Above, in Fig. 1.6, can be found an illustration of the first few of these functions and also an approximation of the graph of Q itself.

Denjoy demonstrated that the function Q is given by the following formula:

$$Q([x_1, x_2, x_3, \ldots]) = -2 \sum_{k=1}^{\infty} (-1)^k 2^{-\sum_{i=1}^{k} x_i}.$$

Later, Salem derived the most important properties of Q from this formula, including the facts that Q is strictly increasing and *singular* with respect to Lebesgue measure, which means that the derivative of Q is equal to zero, Lebesgue-almost everywhere. The function Q is for this reason referred to as a *slippery Devil's staircase*, a term coined by Gutzwiller and Mandelbrot in [GM88]. The (multifractal) fractal nature of Q is investigated in [KS08b], see also [KS07].

Also, recall that the *distribution function* Δ_μ of a measure μ with support in $[0, 1]$ is defined for each $x \in [0, 1]$ by

$$\Delta_\mu(x) := \mu([0, x)).$$

Note that a distribution function is always non-decreasing and right-continuous (see Theorem 9.1.1 in Dudley [Dud89]) and in the case of the measure μ having no atoms, the function Δ_μ is continuous.

Proposition 1.3.7. *The dynamical systems $([0,1],F)$ and $([0,1],T)$ are topologically conjugate and the conjugating homeomorphism Q is given by*

$$Q([x_1,x_2,x_3,\ldots]) := -2\sum_{k=1}^{\infty}(-1)^k 2^{-\sum_{i=1}^{k}x_i}.$$

That is, the conjugating homeomorphism between the Farey map and the tent map is Minkowski's question-mark function. Moreover, the map Q is equal to the distribution function Δ_{μ_0} of the measure of maximal entropy $\mu_0 := \lambda \circ Q$ for the Farey map.

Remark 1.3.8. The reader unfamiliar with the concept of measures of maximal entropy should not be unduly alarmed by this terminology. For our purposes, it is enough to know that μ_0 assigns mass 2^{-n} to each n-th level Farey cylinder set. For further reading we refer to [Wal82]. Also note that since Q is a homeomorphism the measure $\lambda \circ Q$ is well defined.

Proof of Proposition 1.3.7. We will first show that the map Q is the conjugating homeomorphism from F to the tent system. For this, suppose first that $x \in [0,1/2]$. Then, $Q(x)$ is an element of $[0,1/2]$ and we have that

$$T\left(Q(x)\right) = 2\left(-2\sum_{k=1}^{\infty}(-1)^k 2^{-\sum_{i=1}^{k}x_i}\right) = -2\left(\sum_{k=1}^{\infty}(-1)^k 2^{-(x_1-1)-\sum_{i=2}^{k}x_i}\right)$$

$$= Q\left([x_1-1,x_2,x_3,\ldots]\right) = Q(F(x)).$$

Now, suppose that $x \in (1/2,1]$, that is, $x = [1,x_2,x_3,\ldots]$. Then, it follows that $Q(x) \in (1/2,1]$ and we have that

$$T(Q(x)) = 2 - 2\left(2 \cdot 2^{-1} - 2\sum_{k=2}^{\infty}(-1)^k 2^{-1-\sum_{i=2}^{k}x_i}\right)$$

$$= 2\left(\sum_{k=2}^{\infty}(-1)^k 2^{-\sum_{i=2}^{k}x_i}\right) = Q\left([x_2,x_3,\ldots]\right) = Q(F(x)).$$

It only remains to show that Q is equal to the distribution function of μ_0. Indeed, for each $x \in [0,1]$ we have

$$\Delta_{\mu_0}(x) = \mu_0([0,x]) = \lambda \circ Q([0,x]) = \lambda([Q(0),Q(x)]) = \lambda([0,Q(x)]) = Q(x).$$

This finishes the proof. ☐

In preparation for the next proposition, we recall the definition of a Hölder continuous function.

Definition 1.3.9. A map $S : (X, d) \to (X, d)$ of a metric space (X, d) into itself is said to be *Hölder continuous with exponent* $\kappa > 0$ if there exists a positive constant $C > 0$ such that

$$d(S(x), S(y)) \leq C\, d(x, y)^{\kappa}, \text{ for all } x, y \in X.$$

We will show now that Minkowski's question-mark function, the conjugating homeomorphism between the Farey and tent systems, is Hölder continuous with exponent $\log 2/(2 \log \gamma)$, where we recall that $\gamma := (1 + \sqrt{5})/2$ denotes the golden mean. This result was also originally proved by Salem [Sal43]. First, we give a useful lemma.

Lemma 1.3.10. *For the denominator q_n of the n-th convergent of $x = [x_1, x_2, x_3, \ldots]$, we have that $q_n < \gamma^{x_1 + \cdots + x_n}$, for all $n \in \mathbb{N}$.*

Proof. We will prove this by induction. Certainly, for $p_1/q_1 = 1/x_1$, the inequality $x_1 = q_1 < \gamma^{x_1}$ holds. So, fix $n \in \mathbb{N}$ and suppose that $q_k < \gamma^{x_1 + \cdots + x_k}$ for all $k < n$. Recall that $q_n = x_n q_{n-1} + q_{n-2}$. Therefore,

$$q_n < x_n \gamma^{x_1 + \cdots + x_{n-1}} + \gamma^{x_1 + \cdots + x_{n-2}}.$$

Thus, it suffices to show that

$$x_n \gamma^{x_{n-1}} + 1 \leq \gamma^{x_{n-1} + x_n}, \text{ or, that } x_n + 1/\gamma \leq \gamma^{x_n}.$$

This last inequality becomes the equality $1 + 1/\gamma = \gamma$ when $x_n = 1$; it is also straightforward to check that $2 + 1/\gamma = \gamma^2$. As the function $n \mapsto \gamma^n - n$ is increasing for $n \geq 2$, the proof is finished. $\square$

Proposition 1.3.11. *The map Q is s-Hölder continuous for $s = \log 2/(2 \log \gamma)$, but not for any $s > \log 2/(2 \log \gamma)$.*

Proof. In order to calculate the Hölder exponent of Q, first note that

$$\left| Q(C(x_1, x_2, \ldots, x_n)) \right| = \left| Q([x_1, x_2, \ldots, x_n]) - Q([x_1, x_2, \ldots, x_n, 1]) \right|$$
$$= 2^{-s_n},$$

where we have put, as in the proof of Proposition 1.3.7, $s_n := \sum_{i=1}^{n} x_i$. This can be seen by simply calculating the image of the endpoints of this cylinder, or by noting that every Gauss cylinder $C(x_1, x_2, \ldots, x_n)$ is a s_n-th level Farey cylinder.

Now, we have that

$$\lambda(C(x_1, \ldots, x_n)) = |p_n/q_n - p'_{n+1}/q'_{n+1}|, \text{ where } p'_{n+1}/q'_{n+1} = [x_1, \ldots, x_n, 1].$$

Recalling that $p_n q_{n-1} - p_{n-1} q_n = (-1)^{n+1}$, it follows, in light of Lemma 1.3.10, that

$$\lambda(C(x_1, \ldots, x_n)) = \frac{1}{q_n q'_{n+1}} > \gamma^{-(2s_n+1)}.$$

Rearranging this expression, we obtain that

$$\lambda(C(x_1, \ldots, x_n)) > (2^{\log \gamma / \log 2})^{-(2s_n+1)}$$
$$= 2^{-\log \gamma / \log 2} \cdot (2^{-s_n})^{2 \log \gamma / \log 2}$$
$$= c \cdot \left| Q(C(x_1, \ldots, x_n)) \right|^{2 \log \gamma / \log 2},$$

where $c := 2^{-\log \gamma / \log 2} = \gamma^{-1}$. In other words,

$$\left| Q(C(x_1, \ldots, x_n)) \right| \leq c \cdot \lambda(C(x_1, \ldots, x_n))^{\log 2 / (2 \log \gamma)}.$$

Now, let x and y be two arbitrary different irrational numbers in $[0, 1]$. There must be a first time during the backwards iteration of $[0, 1]$ under the inverse branches of F in which a Farey cylinder set appears between the numbers x and y. Say that this cylinder set appears in the p-th stage of the Farey decomposition. If we iterate one more time, it is clear that there are two $(p + 1)$-th level Farey cylinder sets fully contained in the interval (x, y); moreover, one of these also has to be a Gauss cylinder set. Let this Gauss cylinder set be denoted by $C(z_1, z_2, \ldots, z_k)$, where $\sum_{j=1}^{k} z_j = p + 1$. This leads to the observation that, as $C(z_1, z_2, \ldots, z_k)$ is contained in the interval (x, y), we have

$$|x - y|^{\log 2 / (2 \log \gamma)} > \lambda(C(z_1, z_2, \ldots, z_k))^{\log 2 / (2 \log \gamma)}$$
$$\geq c^{-1} \cdot |Q(C(z_1, z_2, \ldots, z_k))| = c^{-1} \cdot 2^{-(p+1)}.$$

Consider the interval (x, y) again. By construction, it is contained inside two neighbouring $(p - 1)$-th level Farey intervals, and so

$$|Q(x) - Q(y)| < 2^{-(p-1)} + 2^{-(p-1)} = 2^{-(p-2)} = 8 \cdot 2^{-(p+1)}.$$

Combining these observations, we obtain that

$$|Q(x) - Q(y)| \leq 8c \, |x - y|^{\log 2 / (2 \log \gamma)}.$$

Now fix $s > \log(2)/(2 \log(\gamma))$ and let x_n, y_n be the left and right boundary point of the Gauss cylinder $C(1, \ldots, 1)$ of length $n \in \mathbb{N}$. Then there exists a constant $c > 0$ such that on the one hand

$$|x_n - y_n| = \lambda(C(1, \ldots, 1)) = \frac{1}{f_n f_{n+1}} \leq c\gamma^{-2n}.$$

(cf. Exercise 1.6.4). On the other hand we have

$$|Q(x_n) - Q(y_n)| = \lambda(Q(C(1, \ldots, 1))) \geq c^{-1} 2^{-n}.$$

Combining these two observations gives

$$\frac{|Q(x_n) - Q(y_n)|}{|x_n - y_n|^s} \geq c^{-s-1} \frac{2^{-n}}{\gamma^{-2sn}} = c^{-s-1} \exp(n(-\log(2) + 2s\log(\gamma))) \to \infty.$$

This finishes the proof. □

1.4 Two further examples

In this section we will introduce and study the properties of our other main examples. These are two families of dynamical systems, the α-Lüroth and α-Farey systems, which are both indexed by partitions α of the unit interval. We shall first introduce the class of partitions of $[0, 1]$ we are interested in and then define the α-Lüroth map L_α. We then describe the expansion of real numbers that can be derived from this map, again in terms of a Markov partition. Next, we introduce the family of α-Farey maps and develop topological properties for these maps similar to those described in Section 1.3.2 for the Farey map.

1.4.1 The α-Lüroth maps

Let us begin by letting $\alpha := \{A_n : n \in \mathbb{N}\}$ denote a countably infinite partition of the unit interval $[0, 1]$, consisting of non-empty, right-closed and left-open intervals. It is assumed throughout that the elements of α are ordered from right to left, starting from A_1, and that these elements accumulate only at the origin. We will denote the collection of all such partitions of $[0, 1]$ by $\mathcal{A}$. Further, we let a_n denote the Lebesgue measure $\lambda(A_n)$ of $A_n \in \alpha$ and let $t_n := \sum_{k=n}^{\infty} a_k$ denote the Lebesgue measure of the n-th tail of α. It is clear that $t_1 = \sum_{k=1}^{\infty} a_k = 1$ for every partition $\alpha \in \mathcal{A}$.

Definition 1.4.1. For a given partition $\alpha \in \mathcal{A}$, the α-Lüroth map $L_\alpha : [0, 1] \to [0, 1]$ is given by

$$L_\alpha(x) := \begin{cases} (t_n - x)/a_n & \text{for } x \in A_n, \ n \in \mathbb{N}; \\ 0 & \text{if } x = 0. \end{cases}$$

In other words, the map L_α consists of countably many linear branches that send A_n onto $[0, 1)$, for each $n \in \mathbb{N}$.

We now define a Markov partition for the map L_α. Let $\mathcal{E} := \{0\}$ and let $\mathring{\alpha} := \{B_n : n \in \mathbb{N}\}$, where $B_n := \text{Int}(A_n)$ denotes the interior of A_n for each $n \in \mathbb{N}$. It is easy to verify that the collection of sets $\mathring{\alpha}$ constitutes a Markov partition. Indeed, the properties (a) and (b) of Definition 1.2.21 are obviously satisfied and property (c) follows from the observation that for all $n \in \mathbb{N}$ we have $L_\alpha(B_n) = (0, 1)$.

For later use, let us also define the inverse branches of L_α. These are the countable family of maps $L_{\alpha,n} : (0, 1) \to B_n$ defined for each $n \in \mathbb{N}$ by

$$L_{\alpha,n}(x) := t_n - a_n x.$$

In order to construct a coding from this Markov partition, we must first show that it is shrinking (cf. Definition 1.2.24). To do this, we calculate the Lebesgue measure of the intervals that make up the refinements $\mathring{\alpha}^n$ of $\mathring{\alpha}$. First of all, the size of each element $B_n := (t_{n+1}, t_n)$ of $\mathring{\alpha}$ is equal to a_n. The refinement $\mathring{\alpha}^2$ is given by

$$\mathring{\alpha}^2 = \mathring{\alpha} \vee L_\alpha^{-1}(\mathring{\alpha}) = L_\alpha^{-1}(\mathring{\alpha}) = \bigcup_{k \in \mathbb{N}} \bigcup_{n \in \mathbb{N}} L_{\alpha,n}(B_k)$$

$$= \bigcup_{k \in \mathbb{N}} \bigcup_{n \in \mathbb{N}} (t_n - a_n t_k, t_n - a_n t_{k+1}).$$

Thus, the size of an interval $B_{n,k} := (t_n - a_n t_k, t_n - a_n t_{k+1})$ in $\mathring{\alpha}^2$ is equal to $a_n a_k$, since $t_k - t_{k+1} = a_k$. Now, suppose that the endpoints of an interval $B_{\ell_1,\dots,\ell_n}$ in the refinement $\mathring{\alpha}^n$ are given by

$$t_{\ell_1} - a_{\ell_1} t_{\ell_2} + \dots + (-1)^{n-1} a_{\ell_1} \dots a_{\ell_{n-1}} t_{\ell_n}$$

and

$$t_{\ell_1} - a_{\ell_1} t_{\ell_2} + \dots + (-1)^{n-1} a_{\ell_1} \dots a_{\ell_{n-1}} t_{\ell_n+1},$$

so that the size of $B_{\ell_1,\dots,\ell_n}$ is equal to $a_{\ell_1} \dots a_{\ell_n}$. To shorten this notation, let us write $[\ell_1, \dots, \ell_n]_a := t_{\ell_1} - a_{\ell_1} t_{\ell_2} + \dots + (-1)^{n-1} a_{\ell_1} \dots a_{\ell_{n-1}} t_{\ell_n}$. Then,

$$\mathring{\alpha}^{n+1} = L_\alpha^{-1}(\mathring{\alpha}^n) = \bigcup_{k \in \mathbb{N}} \bigcup_{(\ell_1,\dots,\ell_n) \in \mathbb{N}^n} L_{\alpha,k}(B_{\ell_1,\dots,\ell_n})$$

$$= \bigcup_{k \in \mathbb{N}} \bigcup_{(\ell_1,\dots,\ell_n) \in \mathbb{N}^n} (t_k - a_k[\ell_1, \dots, \ell_n]_a, t_k - a_k[\ell_1, \dots, \ell_n + 1]_a)_\pm$$

$$= \bigcup_{(\ell_1,\dots,\ell_n,\ell_{n+1}) \in \mathbb{N}^{n+1}} ([\ell_1, \dots, \ell_{n+1}]_a, [\ell_1, \dots, \ell_{n+1} + 1]_a)_\pm,$$

where we recall that the notation $(\cdot, \cdot)_\pm$ means that the endpoints are not necessarily in the correct order. Thus, the size of an interval $B_{\ell_1,\dots,\ell_{n+1}}$ is equal to $a_{\ell_1} \dots a_{\ell_{n+1}}$. From this, we can deduce that the partition $\mathring{\alpha}$ is shrinking. Since $\sum_{k=1}^\infty a_k = 1$, it follows that there exists (at least) one of the a_k with maximum size. Call it $a_{\max}$. Then, the largest element of $\mathring{\alpha}^n$ has size $(a_{\max})^n$ and, as $0 < a_{\max} < 1$, this clearly tends to zero as n tends to infinity.

Now we can utilise the shrinking Markov partition $\mathring{\alpha}$ to obtain a coding for all the points in the set $[0, 1] \setminus \bigcup_{n=0}^\infty L_\alpha^{-n}(0)$. Exactly as was the case for the Gauss map before, if x lies in this set, we find a sequence $(\ell_1, \ell_2, \dots) \in \mathbb{N}^\mathbb{N}$ such that

$x \in \bigcap_{n=0}^{\infty} \overline{B_{\ell_1} \cap L_\alpha^{-1} B_{\ell_2} \cap \cdots \cap L_\alpha^{-n} B_{\ell_{n+1}}}$. Thus, $x \in B_{\ell_1}$ and therefore,

$$L_\alpha(x) = (t_{\ell_1} - x)/a_{\ell_1}.$$

So, $x = t_{\ell_1} - a_{\ell_1} L_\alpha(x)$. Then, $L_\alpha(x) \in B_{\ell_2}$ and a similar calculation leads us to the identity

$$x = t_{\ell_1} - a_{\ell_1} t_{\ell_2} + a_{\ell_1} a_{\ell_2} L_\alpha^2(x).$$

We can continue this calculation indefinitely to obtain an alternating series expansion of each $x \in [0,1] \setminus \bigcup_{n=0}^{\infty} L_\alpha^{-n}(0)$, which is given by

$$x = t_{\ell_1} + \sum_{n=2}^{\infty} (-1)^{n-1} \left(\prod_{i<n} a_{\ell_i} \right) t_{\ell_n} = t_{\ell_1} - a_{\ell_1} t_{\ell_2} + a_{\ell_1} a_{\ell_2} t_{\ell_3} - \cdots.$$

This will be called the α-Lüroth expansion of the point x. To shorten the notation, we denote these infinite series expansions by $x = [\ell_1, \ell_2, \ell_3, \ldots]_\alpha$.

Notice that the infinite α-Lüroth expansions match with the finite ones we obtained above whilst calculating the endpoints of the intervals of the refined partitions $\mathring{\alpha}^n$. The main difference is that each infinite expansion is unique (as in the case of infinite continued fractions), whereas the finite ones can be written in either of the two ways:

$$[\ell_1, \ell_2, \ldots, \ell_k]_\alpha := t_{\ell_1} - a_{\ell_1} t_{\ell_2} + \cdots + (-1)^{k-1} a_{\ell_1} \ldots a_{\ell_{k-1}} t_{\ell_k}$$

and

$$[\ell_1, \ell_2, \ldots, \ell_k - 1, 1]_\alpha = t_{\ell_1} - a_{\ell_1} t_{\ell_2} + \cdots + (-1)^k a_{\ell_1} \ldots a_{\ell_{k-1}} a_{\ell_k - 1} t_1,$$

where we assume that $\ell_k > 1$.

By analogy with continued fractions, for which a number is rational if and only if it has a finite continued fraction expansion, we say that $x \in [0,1]$ is an α-rational number when x has a finite α-Lüroth expansion - that is, whenever x is a pre-image of 0 under the map L_α - and say that x is an α-irrational number otherwise. We denote the set of α-rational numbers by $\mathbb{Q}_\alpha$ and the set of α-irrational numbers by $\mathbb{I}_\alpha$. The set $\mathbb{Q}_\alpha$ is, of course, a countable dense set in $[0,1]$. The reader should also notice that the α-rationals are not necessarily equal to actual rational numbers, unless the partition α is chosen to consist solely of intervals with rational endpoints.

Example 1.4.2.
(a) Define the *harmonic partition* α_H by setting

$$\alpha_H := \left\{ A_n := \left(\frac{1}{n+1}, \frac{1}{n} \right] : n \geq 1 \right\}.$$

The map $L_{\alpha_H} : [0, 1] \to [0, 1]$ is then given by

$$L_{\alpha_H}(x) := \begin{cases} -n(n+1)x + (n+1) & \text{for } x \in \left(\dfrac{1}{n+1}, \dfrac{1}{n}\right]; \\ 0 & \text{for } x = 0. \end{cases}$$

This map can be found in the literature where it is often referred to as the *alternating Lüroth map*. For references and more on the historical background to this, see Section 1.5.

With respect to the map L_{α_H}, in exactly the way outlined above using the Markov partition $\mathring{\alpha}_H$, the corresponding series expansion of some arbitrary $x \in [0, 1]$ turns out to be

$$x = \sum_{n=1}^{\infty} \left((-1)^{n-1}(\ell_n + 1) \prod_{k=1}^{n}(\ell_k(\ell_k + 1))^{-1} \right)$$

$$= \frac{1}{\ell_1} - \frac{1}{\ell_1(\ell_1 + 1)\ell_2} + \frac{1}{\ell_1(\ell_1 + 1)\ell_2(\ell_2 + 1)\ell_3} - \cdots,$$

where each $\ell_n \in \mathbb{N}$ and the expansion can, as usual, be finite or infinite.

(b) Define the partition $\alpha_D := \{(1/2^n, 1/2^{n-1}] : n \in \mathbb{N}\}$. We will refer to α_D as the *dyadic partition*. For this partition, we obtain the map L_{α_D}, which is given by

$$L_{\alpha_D}(x) := \begin{cases} 2 - 2^n x & \text{for } x \in (1/2^n, 1/2^{n-1}]; \\ 0 & \text{for } x = 0. \end{cases}$$

Remark 1.4.3.

1. The name "α-Lüroth" for these maps is in honour of the German mathematician J. Lüroth, for his 1883 paper [Lür83] which develops a particular series expansion of real numbers which is related to the expansions derived above. For more details, we refer once more to Section 1.5.

2. Note that the α-Lüroth expansion is a particular type of *generalised Lüroth series*, a concept which was introduced by Barrionuevo *et al.* in [BBDK96] (also see [DK02]).

Before going any further, let us describe the action of the map L_α on the expansions it generates. For each $x = [\ell_1, \ell_2, \ell_3, \ldots]_\alpha$, we have, since $x \in A_{\ell_1}$, that

$$L_\alpha(x) = (t_{\ell_1} - x)/a_{\ell_1} = (t_{\ell_1} - (t_{\ell_1} - a_{\ell_1}t_{\ell_2} + a_{\ell_1}a_{\ell_2}t_{\ell_3} + \ldots))/a_{\ell_1}$$

$$= t_{\ell_2} + a_{\ell_2}t_{\ell_3} + \ldots = [\ell_2, \ell_3, \ell_4, \ldots]_\alpha.$$

This shows that L_α, just like the Gauss map, can be thought of as acting as the shift map on the space $\mathbb{N}^{\mathbb{N}}$, at least for those points in $[0, 1]$ with infinite α-Lüroth expansions. That is, $L_\alpha : \mathbb{I}_\alpha \to \mathbb{I}_\alpha$ and $\sigma : \mathbb{N}^{\mathbb{N}} \to \mathbb{N}^{\mathbb{N}}$ are topologically conjugate via the conjugacy map $h : \mathbb{N}^{\mathbb{N}} \to \mathbb{I}_\alpha$ given by $h(\ell_1\ell_2\ell_3 \ldots) = [\ell_1, \ell_2, \ell_3, \ldots]_\alpha$.

For each $x = [\ell_1, \ell_2, \ell_3, \ldots]_\alpha \in [0, 1]$, just as was done for the continued fraction expansion, if we truncate the α-Lüroth expansion of x after k entries, then we obtain the k-th α-Lüroth convergent of x, that is, for each $k \in \mathbb{N}$ we obtain the finite α-Lüroth expansion $r_k^{(\alpha)}(x)$, given by

$$r_k^{(\alpha)}(x) := [\ell_1, \ldots, \ell_k]_\alpha = t_{\ell_1} - a_{\ell_1} t_{\ell_2} + \cdots + (-1)^{k-1} a_{\ell_1} \cdots a_{\ell_{k-1}} t_{\ell_k}.$$

The behaviour of these convergents is exactly like those of the continued fraction convergents, as shown in the following proposition.

Proposition 1.4.4. *Let $x = [\ell_1, \ell_2, \ell_3, \ldots]_\alpha \in \mathbb{I}_\alpha$. Then, the sequence of α-Lüroth convergents of x satisfies the following four properties.*
(a) *The sequence $\left(r_{2n}^{(\alpha)}(x)\right)_{n \geq 1}$ of even convergents is increasing.*
(b) *The sequence $\left(r_{2n-1}^{(\alpha)}(x)\right)_{n \geq 1}$ of odd convergents is decreasing.*
(c) *Every convergent of odd order is greater than every convergent of even order.*
(d) $\lim_{n \to \infty} \left| r_{n+1}^{(\alpha)}(x) - r_n^{(\alpha)}(x) \right| = 0.$

Proof. The proof is very similar to that of Proposition 1.1.3 and, as such, is left as an exercise. □

Definition 1.4.5. For each k-tuple $(\ell_1, \ldots, \ell_k)$ of positive integers, define the α-*Lüroth cylinder set* $C_\alpha(\ell_1, \ldots, \ell_k)$ associated with the α-Lüroth expansion by

$$C_\alpha(\ell_1, \ldots, \ell_k) := \{[y_1, y_2, \ldots]_\alpha : y_i = \ell_i \text{ for } 1 \leq i \leq k\}.$$

Observe once again that these cylinder sets coincide up to sets of measure zero with the elements of the refinements $\mathring{\alpha}^n$ of the Markov partition $\mathring{\alpha}$.

1.4.2 The α-Farey maps

Let us now introduce a second family of maps, indexed by the same collection $\mathcal{A}$ of partitions of $[0, 1]$ as were used in the definition of L_α. We will soon see that these new maps are related to the maps L_α in the same way the Farey map is related to the Gauss map.

Definition 1.4.6. For a given partition $\alpha := \{A_n : n \in \mathbb{N}\} \in \mathcal{A}$, the α-*Farey map* $F_\alpha : [0, 1] \to [0, 1]$ is defined by

$$F_\alpha(x) := \begin{cases} (1 - x)/a_1 & \text{if } x \in A_1; \\ a_{n-1}(x - t_{n+1})/a_n + t_n & \text{if } x \in A_n, \text{ for } n \geq 2; \\ 0 & \text{if } x = 0. \end{cases}$$

Although the formula looks a bit cryptic, all that the transformation F_α does is to map the set A_1 linearly onto the interval $[0, 1)$ and, for each $n \geq 2$, map the interval A_n linearly onto the interval A_{n-1}. In particular, notice that $F_\alpha|_{A_1} = L_\alpha|_{A_1}$. The action of F_α on each point $x = [\ell_1, \ell_2, \ldots]_\alpha \in [0, 1]$ is given by

$$F_\alpha(x) := \begin{cases} [\ell_2, \ell_3, \ldots]_\alpha & \text{for } \ell_1 = 1; \\ [\ell_1 - 1, \ell_2, \ell_3, \ldots]_\alpha & \text{for } \ell_1 \geq 2. \end{cases}$$

Notice that the map F_α acts on the α-Lüroth expansion of x in precisely the same way as the Farey map acts on the continued fraction expansion of each point $x \in [0, 1]$.

Definition 1.4.7. Let the two inverse branches of the map F_α be denoted by

$$F_{\alpha,0} : [0, 1] \rightarrow [0, t_2] \text{ and } F_{\alpha,1} : [0, 1] \rightarrow [t_2, 1].$$

With the convention that $F_{\alpha,0}(0) = 0$, these two branches are given by

$$F_{\alpha,0}(x) := \frac{a_{n+1}}{a_n}(x - t_{n+1}) + t_{n+2}, \text{ for } x \in A_n, \ n \geq 1$$

and

$$F_{\alpha,1}(x) := 1 - a_1 x, \text{ for } x \in [0, 1].$$

Note that $F_{\alpha,0}$ maps the interval A_n into the interval A_{n+1}, for each $n \in \mathbb{N}$.

Example 1.4.8.
(a) For the harmonic partition $\alpha_H := \{A_n := (1/(n+1), 1/n] : n \in \mathbb{N}\}$, we obtain the α_H-Farey map F_{α_H}, which is given explicitly by

$$F_{\alpha_H}(x) := \begin{cases} 2 - 2x & \text{for } x \in (1/2, 1]; \\ \dfrac{n+1}{n-1}x - \dfrac{1}{n(n-1)} & \text{for } x \in (1/(n+1), 1/n]. \end{cases}$$

The graphs of the maps L_{α_H} and F_{α_H} are shown in Fig. 1.7
(b) Consider again the dyadic partition $\alpha_D := \{(1/2^n, 1/2^{n-1}] : n \in \mathbb{N}\}$. In this case the map F_{α_D} coincides with the tent map. To see this, it is enough to note that for each $n \in \mathbb{N}$ we have that $a_n = 2^{-n}$ and $t_n = 2^{-(n-1)}$. The graphs of the maps L_{α_D} and F_{α_D} are shown in Fig. 1.8.

We now show that the relationship between the maps L_α and F_α is exactly the same as the relationship between the maps G and F, that is, L_α is a jump transformation of F_α. More precisely, we make the following definition.

Definition 1.4.9. Let the map $\rho_\alpha : (0, 1] \rightarrow \mathbb{N} \cup \{0\}$ be defined by setting

$$\rho_\alpha(x) := \inf\{n \geq 0 : F_\alpha^n(x) \in A_1\}.$$

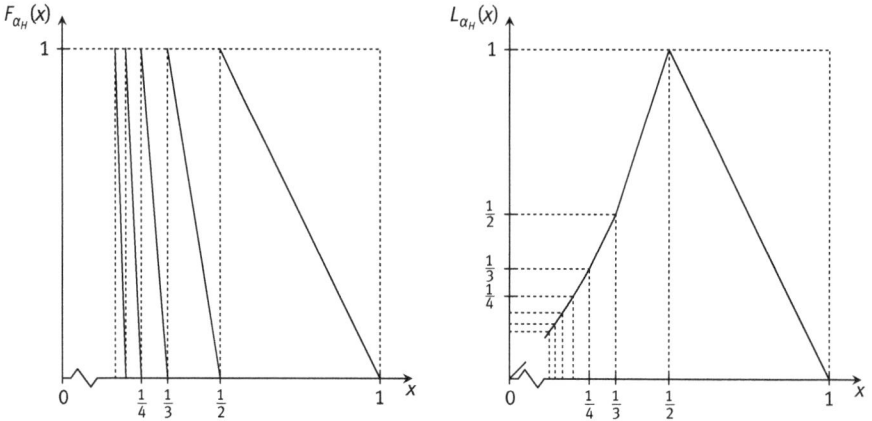

Fig. 1.7. The α_H-Lüroth and α_H-Farey map, where $t_n = 1/n$, $n \in \mathbb{N}$.

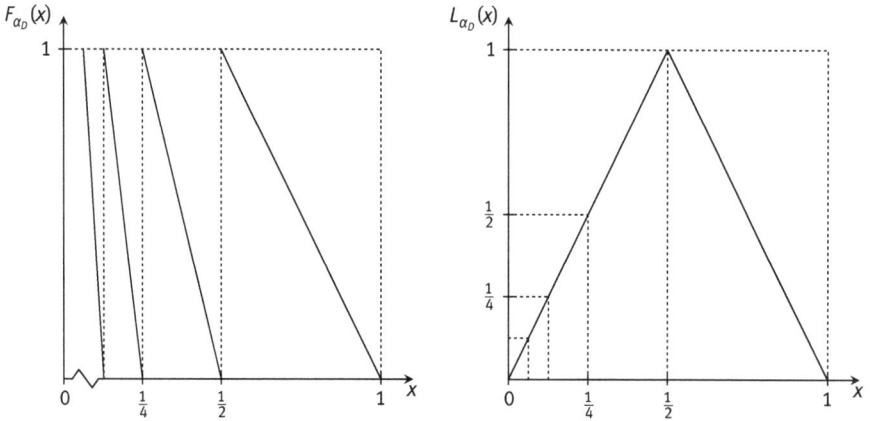

Fig. 1.8. The α_D-Lüroth and α_D-Farey map, which coincides with the tent map T, where $t_n = (1/2)^{n-1}$, $n \in \mathbb{N}$.

Notice that the map ρ_α is finite everywhere on $(0, 1]$. Then, let the map $F_\alpha^* : [0, 1] \to [0, 1]$ be defined by

$$F_\alpha^*(x) := \begin{cases} F_\alpha^{\rho_\alpha(x)+1}(x) & \text{if } x \neq 0; \\ 0 & \text{if } x = 0. \end{cases}$$

The map F_α^* is said to be the *jump transformation on* A_1 of F_α.

We then obtain the following result. Note that the proof can be copied line by line from the proof of the corresponding result for the Farey and Gauss maps, so we omit it here.

Lemma 1.4.10. *The jump transformation F_α^* of F_α coincides with the α-Lüroth map L_α.*

Proof. This follows in precisely the same way as Lemma 1.3.3. □

Let us now describe how to construct a Markov partition for the α-Farey map from the partition α and use it to obtain another coding of the points in $[0, 1]$ from the map F_α. The partition given by the two open sets $\{B_0 := \mathrm{Int}(A_1), B_1 := (0, 1]\setminus\overline{A_1}\}$ is a Markov partition for F_α. Each α-irrational number in $[0, 1]$ has an infinite coding $x =: \langle x_1, x_2, \ldots\rangle_\alpha$ with $(x_1, x_2, \ldots) \in \{0, 1\}^{\mathbb{N}}$ such that $x_k = 1$ if and only if $F_\alpha^{k-1}(x) \in B_1$ for each $k \in \mathbb{N}$. This coding will be referred to as the α-*Farey expansion* or the α-*Farey coding*. (Here we are skipping over the details, but this coding is obtained in precisely the way that the Farey coding was obtained in Section 1.3.) The α-Farey coding is related to the α-Lüroth coding in exactly the same way as the Farey coding is related to the continued fraction expansion, namely, if an α-irrational number $x \in [0, 1]$ has α-Lüroth coding given by $x = [\ell_1, \ell_2, \ell_3, \ldots]_\alpha$, then the α-Farey coding of x is given by $x := \langle 0^{\ell_1-1}, 1, 0^{\ell_2-1}, 1, 0^{\ell_3-1}, 1, \ldots\rangle_\alpha$, where we recall that 0^n denotes the sequence of $n \in \mathbb{N}$ consecutive appearances of the symbol 0 and 0^0 is understood to mean the appearance of no zeros between two consecutive 1s. For each α-rational number $x = [\ell_1, \ell_2, \ldots, \ell_k]_\alpha$, one immediately verifies that this number has an α-Farey coding given by either

$$x = \langle 0^{\ell_1-1}, 1, 0^{\ell_2-1}, 1, \ldots, 0^{\ell_k-1}, 1, 0, 0, 0, \ldots\rangle_\alpha$$

or

$$x = \langle 0^{\ell_1-1}, 1, 0^{\ell_2-1}, 1, \ldots, 0^{\ell_k-2}, 1, 1, 0, 0, 0, \ldots\rangle_\alpha.$$

Recall that F_α acts on $x = [\ell_1, \ell_2, \ldots]_\alpha$ in the following way:

$$F_\alpha(x) := \begin{cases} [\ell_1 - 1, \ell_2, \ell_3, \ldots]_\alpha & \text{for } \ell_1 \geq 2; \\ [\ell_2, \ell_3, \ldots]_\alpha & \text{for } \ell_1 = 1. \end{cases}$$

In particular, this means that if we instead write x in its α-Farey coding, that is, $x = \langle x_1, x_2, x_3, \ldots\rangle_\alpha$, then

$$F_\alpha(x) := \langle x_2, x_3, x_4, \ldots\rangle_\alpha.$$

Therefore, the map $F_\alpha : [0, 1] \to [0, 1]$ is a factor of the shift map σ on the shift space $\{0, 1\}^{\mathbb{N}}$, via the factor map $h : \{0, 1\}^{\mathbb{N}} \to [0, 1]$ defined by $h((x_1, x_2, x_3, \ldots)) := \langle x_1, x_2, x_3, \ldots\rangle_\alpha$.

Let us now define the cylinder sets associated with the map F_α. These once more coincide with the refinements of the Markov partition given above for F_α.

Definition 1.4.11. For each n-tuple $(x_1, \ldots, x_n)$ of positive integers, define the α-*Farey cylinder set* $\widehat{C}_\alpha(x_1, \ldots, x_n)$ by setting

$$\widehat{C}_\alpha(x_1, \ldots, x_n) := \{\langle y_1, y_2, \ldots \rangle_\alpha : y_k = x_k, \text{ for } 1 \leq k \leq n\}.$$

By analogy with the Farey decomposition described after Definition 1.3.4, we will call the set of cylinder sets $\{\widehat{C}_\alpha(x_1, \ldots, x_n) : (x_1, \ldots, x_n) \in \{0, 1\}^n\}$ the n-*th level α-Farey decomposition*. Observe that we have the relation $\widehat{C}_\alpha(x_1, \ldots, x_n) = F_{\alpha, x_1} \circ \cdots \circ F_{\alpha, x_n}([0, 1])$.

Notice that every α-Lüroth cylinder set is also an α-Farey cylinder set, whereas the converse of this statement is not true. The precise description of the correspondence is that any α-Farey cylinder set which has the form $\widehat{C}_\alpha(0^{\ell_1 - 1}, 1, \ldots, 0^{\ell_k - 1}, 1)$ coincides with the α-Lüroth cylinder set $C_\alpha(\ell_1, \ldots, \ell_k)$, but if an α-Farey cylinder set is defined by a finite word ending in the symbol 0, then it cannot be translated to a single α-Lüroth cylinder set. However, we do have the relation

$$\widehat{C}_\alpha(0^{\ell_1 - 1}, 1, 0^{\ell_2 - 1}, 1, \ldots, 0^{\ell_k - 1}, 1, 0^m) = \bigcup_{n \geq m + 1} C_\alpha(\ell_1, \ell_2, \ldots, \ell_k, n).$$

It therefore follows that for the Lebesgue measure of this interval we have that

$$\lambda(\widehat{C}_\alpha(0^{\ell_1 - 1}, 1, 0^{\ell_2 - 1}, 1, \ldots, 0^{\ell_k - 1}, 1, 0^m)) = \sum_{n \geq m + 1} \lambda(C_\alpha(\ell_1, \ell_2, \ldots, \ell_k, n))$$

$$= a_{\ell_1} a_{\ell_2} \cdots a_{\ell_k} t_{m+1}.$$

In addition, we can identify the endpoints of each α-Farey cylinder set. If we consider the set $\widehat{C}_\alpha(0^{\ell_1 - 1}, 1, \ldots, 0^{\ell_k - 1}, 1)$, then we already know the endpoints of this interval (since it is also equal to an α-Lüroth cylinder set). On the other hand, the endpoints of the set $\widehat{C}_\alpha(0^{\ell_1 - 1}, 1, 0^{\ell_2 - 1}, 1, \ldots, 0^{\ell_k - 1}, 1, 0^m)$ are given by $[\ell_1, \ldots, \ell_k, m + 1]_\alpha$ and $[\ell_1, \ldots, \ell_k]_\alpha$.

1.4.3 Topological properties of F_α

Let us now consider the topological properties of the maps F_α. Perhaps by now the reader will not be surprised to learn that they are essentially the same as the topological properties of the Farey map F. Again, the proofs can be closely modelled after the proofs of the corresponding results for the Farey map and so we leave many of them as exercises.

Before stating the first proposition, we remind the reader that the measure of maximal entropy μ_α for the system F_α is the measure that assigns mass 2^{-n} to each n-th level α-Farey cylinder set, for each $n \in \mathbb{N}$.

Proposition 1.4.12. *The dynamical systems* $([0, 1], F_\alpha)$ *and* $([0, 1], T)$ *are to-pologically conjugate and the conjugating homeomorphism is given, for each* $x = [\ell_1, \ell_2, \ell_3, \ldots]_\alpha$, *by*

$$\theta_\alpha(x) := -2 \sum_{k=1}^{\infty} (-1)^k 2^{-\sum_{i=1}^{k} \ell_i}.$$

Moreover, the map θ_α *is equal to the distribution function of the measure of maximal entropy* μ_α *for the* α-*Farey map.*

Proof. See Exercise 1.6.17. □

Notice that the only difference between the map θ_α, for a given partition α, and Minkowski's question-mark function Q is that the summands in the power of 2 in the latter are the elements of the continued fraction expansion of the point x, whereas in the formula for the function θ_α the elements of the α-Lüroth expansion turn up. Each function θ_α is continuous and strictly increasing. It can also be shown that each of the functions θ_α (with the obvious, trivial exception of the function θ_{α_D} which simply maps the tent map to the tent map), are singular with respect to the Lebesgue measure (for more on these functions, see [Mun14]).

Our next aim is to determine the Hölder exponent and the sub-Hölder exponent of the map θ_α, for an arbitrary partition α. Recall that the definition of a Hölder continuous function was given in Definition 1.3.9. In a similar vein, we say that a map $S : X \to X$ of a metric space (X, d) into itself is *sub-Hölder continuous with exponent* $\kappa > 0$ if there exists a positive constant C such that

$$d(S(x), S(y)) \geq C d(x, y)^\kappa, \text{ for all } x, y \in X.$$

In order to determine the Hölder and sub-Hölder exponents of θ_α, let us first define $\kappa(n) := -n \log 2 / (\log a_n)$ and set

$$\kappa_+ := \inf \{\kappa(n) : n \in \mathbb{N}\} \text{ and } \kappa_- := \sup \{\kappa(n) : n \in \mathbb{N}\}.$$

Proposition 1.4.13. *We have that the map* θ_α *is* κ_+-*Hölder continuous and, provided that* κ_- *is finite,* κ_--*sub-Hölder continuous.*

Proof. In order to calculate the Hölder exponent of θ_α, first note that

$$|\theta_\alpha(C_\alpha(\ell_1, \ell_2, \ldots, \ell_k))| = |\theta_\alpha([\ell_1, \ell_2, \ldots, \ell_k]_\alpha) - \theta_\alpha([\ell_1, \ell_2, \ldots, \ell_k + 1]_\alpha)|$$
$$= 2^{-\sum_{j=1}^{k} \ell_j}.$$

This can be seen by simply calculating the image of the endpoints of this cylinder, or by noting that every α-Lüroth cylinder set $C_\alpha(\ell_1, \ell_2, \ldots, \ell_k)$ is an n-th level α-Farey cylinder set, where $n = \sum_{j=1}^{k} \ell_j$.

Suppose first that κ_+ is non-zero. In this case we have that

$$\lambda(C_\alpha(\ell_1, \ell_2, \ldots, \ell_k)) = \prod_{i=1}^{k} a_{\ell_i} = \prod_{i=1}^{k} 2^{-\ell_i/\kappa(\ell_i)} \geq \left(\prod_{i=1}^{k} 2^{-\ell_i} \right)^{1/\kappa_+}$$

$$= \left(2^{-\sum_{i=1}^{k} \ell_i} \right)^{1/\kappa_+} = |\theta_\alpha(C_\alpha(\ell_1, \ell_2, \ldots, \ell_k))|^{1/\kappa_+}.$$

Or, in other words,

$$|\theta_\alpha(C_\alpha(\ell_1, \ell_2, \ldots, \ell_k))| \leq \lambda(C_\alpha(\ell_1, \ell_2, \ldots, \ell_k))^{\kappa_+}.$$

From this point, the proof that θ_α is κ_+-Hölder continuous is completed in precisely the same way as the proof that Q is $\log 2/(2 \log \gamma)$-Hölder continuous and we leave the details to the reader.

Suppose now that κ_+ is equal to zero. Then, we have that for each $q \in \mathbb{N}$ there exists $m_0 \in \mathbb{N}$ with the property that for every $m \geq m_0$,

$$\kappa(m) = \frac{m \log 2}{-\log a_m} < \frac{1}{q}, \quad \text{or, equivalently,} \quad a_m < 2^{-qm}.$$

So we have that the sequence of partition elements are eventually exponentially decaying, and hence, the Hölder exponent of the map θ_α is necessarily equal to zero.

It remains to show that the map θ_α is κ_--sub-Hölder continuous. Suppose that κ_- is finite (otherwise the definition of sub-Hölder continuity makes no real sense). Similarly to the κ_+ case, we obtain the inequality

$$|\theta_\alpha(C_\alpha(\ell_1, \ldots, \ell_k))|^{1/\kappa_-} \geq \lambda(C_\alpha(\ell_1, \ldots, \ell_k)).$$

The proof can be completed analogously to the proof of the Hölder continuity of θ_α from this point on and the details are left as an exercise for the reader (see Exercise 1.6.19). □

Example 1.4.14. For the conjugacy map θ_{α_H} between the map F_{α_H}, arising from the harmonic partition, and the tent map T, we have that θ_{α_H} is $\log 4/\log 6$-Hölder continuous. To show this, first observe that

$$\kappa(1) = \frac{-\log 2}{\log 1/2} = 1 > \frac{2 \log 2}{\log 6} = \kappa(2) \quad \text{and} \quad \frac{2 \log 2}{\log 6} < \frac{3 \log 2}{\log 12} = \kappa(3).$$

Then, since $6^n > (n^2 + n)^2$ for $n \geq 3$, we have that for all $n \geq 3$,

$$\kappa(n) = \frac{n \log 2}{\log(n(n+1))} > \frac{2 \log 2}{\log 6} = \kappa(2).$$

Concerning the existence of a constant κ for which the map θ_{α_H} is κ-sub-Hölder continuous, recall that this means we have to find $\kappa > 0$ such that for all $x, y \in [0, 1]$,

$$|\theta_{\alpha_H}(x) - \theta_{\alpha_H}(y)| \gg |x - y|^{\kappa}.$$

In particular, this inequality has to be satisfied for $x = 0$ and y given by $[n]_{\alpha_H} = 1/n$ successively, for each $n \in \mathbb{N}$. But here we have that $|\theta_{\alpha_H}(0) - \theta_{\alpha_H}(1/n)| = 2^{-(n-1)}$ and $|0 - 1/n| = 1/n$, which implies that there can be no such κ.

Notice that, in line with the fact that there is no sub-Hölder continuity in this case, we also have that κ_- is infinite.

Remark 1.4.15. The reasoning given above for why the map θ_{α_H} fails to be sub-Hölder continuous also works for the map Q (the conjugacy map between the Farey and tent maps). In other words, there is no positive constant κ such that the map Q is κ-sub-Hölder continuous.

1.4.4 Expanding and expansive partitions

Let us now introduce some particular classes of partitions that will be useful in the chapters that follow, particularly in Chapter 3. Before beginning this task, we first recall the definition of a slowly varying function.

Definition 1.4.16. A measurable function $\psi : \mathbb{R}^+ \to \mathbb{R}^+$ is said to be *slowly varying* if

$$\lim_{x \to \infty} \frac{\psi(xy)}{\psi(x)} = 1, \text{ for all } y > 0.$$

In the following proposition, we list some of the useful properties that slowly varying functions satisfy. From this list, it should be clear that the idea behind a slowly varying function is that it behaves like a logarithmic function.

Proposition 1.4.17. *Let* $\psi, \varphi : \mathbb{R}^+ \to \mathbb{R}^+$ *be two slowly varying functions. Then the following three statements hold:*
(a) For any $\varepsilon > 0$, *we have that*

$$\lim_{x \to \infty} x^{\varepsilon} \cdot \psi(x) = \infty \text{ and } \lim_{x \to \infty} x^{-\varepsilon} \cdot \psi(x) = 0.$$

(b) $\lim_{x \to \infty} \dfrac{\log(\psi(x))}{\log(x)} = 0.$
(c) For any $-\infty < a < \infty$, *the functions* ψ^a, $\psi \cdot \varphi$ *and* $\psi + \varphi$ *are all slowly varying.*

Proof. See the book by Seneta [Sen76]. $\square$

Definition 1.4.18. Let $\alpha := \{A_n : n \in \mathbb{N}\} \in \mathcal{A}$. Then:

(a) The partition α is said to be *expanding* provided that

$$\lim_{n \to \infty} \frac{t_n}{t_{n+1}} = \rho, \quad \text{for some } \rho > 1.$$

(b) The partition α is said to be *expansive of exponent* $\theta \geq 0$ if the tails of the partition satisfy the power law

$$t_n = \psi(n) \cdot n^{-\theta},$$

where $\psi : \mathbb{R}^+ \to \mathbb{R}^+$ is a slowly varying function.

(c) A partition α is said to be of *finite type* if for the sequence of tails t_n of α, we have that $\sum_{n=1}^{\infty} t_n$ converges. Otherwise, α is said to be of *infinite type*.

Notice that if α is expanding, one immediately verifies that α is of finite type. This can be seen, for instance, by applying the ratio test for series convergence. The next proposition describes the situation for expansive partitions.

Proposition 1.4.19. *Suppose that α is expansive of exponent $\theta \geq 0$. Then we have the following classification:*

(a) *If $\theta \in [0, 1)$, then α is of infinite type.*

(b) *If $\theta > 1$, then α is of finite type.*

(c) *If $\theta = 1$, then α can be either of finite or infinite type.*

Proof. Suppose first that α is expansive of exponent $\theta \in [0, 1)$. Then, by Proposition 1.4.17, for all $\varepsilon > 0$ there exists $n_0 \in \mathbb{N}$ such that if $n \geq n_0$, then we have that $\psi(n) \geq n^{-\varepsilon}$. Let $\varepsilon > 0$ be sufficiently small such that $\theta + \varepsilon \in (0, 1)$. Then,

$$\sum_{n=1}^{\infty} t_n = \sum_{n=1}^{n_0-1} \psi(n) \cdot n^{-\theta} + \sum_{n=n_0}^{\infty} \psi(n) \cdot n^{-\theta} \geq \sum_{n=n_0}^{\infty} n^{-(\theta+\varepsilon)} \geq \sum_{n=n_0}^{\infty} n^{-1}.$$

Consequently, $\sum_{n=1}^{\infty} t_n = \infty$ and α is of infinite type. Now suppose that α is expansive of exponent $\theta > 1$. Then, again by Proposition 1.4.17, for all $\varepsilon > 0$ there exists $n_0 \in \mathbb{N}$ such that if $n \geq n_0$, then we have that $\psi(n) \leq n^{\varepsilon}$. For $\varepsilon > 0$ small enough such that $\theta - \varepsilon > 1$, we then have that

$$\sum_{n=n_0}^{\infty} t_n = \sum_{n=n_0}^{\infty} \psi(n) \cdot n^{-\theta} \leq \sum_{n=n_0}^{\infty} n^{-(\theta-\varepsilon)} < \infty.$$

Therefore, in this case, the partition α is of finite type. It only remains to prove the third assertion, which can be done by considering the following two examples. First, let $t_1 := 1$ and for each $n \geq 2$, let $t_n := (n \log n)^{-1}$. The partition α defined in such a way is clearly expansive of exponent 1. For this partition, we have that

$$\sum_{n=1}^{\infty} t_n = 1 + \sum_{n=2}^{\infty} \frac{1}{n\log n},$$

which diverges. So, in this first case, the partition is of infinite type. On the other hand, if now we define a partition by setting $t_1 := 1$ and $t_n := n^{-1} \cdot (\log n)^{-2}$ for $n \geq 2$, we obtain that

$$\sum_{n=1}^{\infty} t_n = 1 + \sum_{n=2}^{\infty} \frac{1}{n(\log n)^2},$$

which is a convergent series, so in this case we have that the partition is of finite type. This finishes the proof. □

Fig. 1.9 illustrates two α-Farey maps with α expansive. The graph on the left-hand side has α with exponent $\theta = 2$, so satisfies the condition of the second part of Proposition 1.4.19. The graph on the right-hand side has α with exponent $\theta = 1/2$, so it satisfies the condition given in the first part of Proposition 1.4.19.

1.4.5 Metrical Diophantine-like results for the α-Lüroth expansion

In this section, we consider some easily-obtained results for the α-Lüroth expansion which are analogous to the metrical Diophantine results already given above

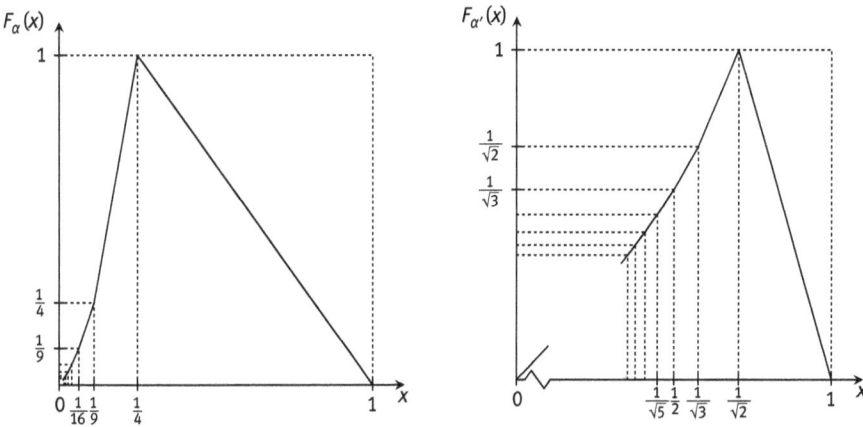

Fig. 1.9. The graphs of two α-Farey maps with α expansive. The partition on the left is of finite type with α given by $t_n := 1/n^2$, $n \in \mathbb{N}$ and the partition on the right is of infinite type with α' given by $t_n := 1/\sqrt{n}$, $n \in \mathbb{N}$.

for the continued fraction expansion. We will first consider the equivalent of the badly-approximable numbers.

Definition 1.4.20. For each $N \in \mathbb{N}$, let the set $\mathcal{B}_{\alpha,N}$ be defined by

$$\mathcal{B}_{\alpha,N} := \{x = [\ell_1, \ell_2, \ldots]_\alpha \in \mathbb{I}_\alpha : \ell_n \leq N \text{ for all sufficiently large } n \in \mathbb{N}\}$$

and set

$$\mathcal{B}_\alpha := \bigcup_{N \in \mathbb{N}} \mathcal{B}_{\alpha,N}.$$

The set $\mathcal{B}_\alpha$ will be referred to as the set of *badly α-approximable numbers*.

Lemma 1.4.21.

$$\lambda(\mathcal{B}_\alpha) = 0.$$

Proof. First notice that we can write the set of badly α-approximable numbers in the following way:

$$\mathcal{B}_\alpha = \bigcup_{N \in \mathbb{N}} \mathcal{A}_{\alpha,N},$$

where

$$\mathcal{A}_{\alpha,N} := \{x = [\ell_1, \ell_2, \ldots]_\alpha \in \mathbb{I}_\alpha : \ell_k \leq N \text{ for all } k \in \mathbb{N}\}.$$

Also, for each $N, n \in \mathbb{N}$, define

$$\mathcal{A}_{\alpha,N}^{(n)} := \{x = [\ell_1, \ell_2, \ldots]_\alpha \in \mathbb{I}_\alpha : \ell_k \leq N \text{ for all } 1 \leq k \leq n\}.$$

It is clear that $\mathcal{A}_{\alpha,N} \subseteq \mathcal{A}_{\alpha,N}^{(n)}$ and further that $\mathcal{A}_{\alpha,N}^{(n+1)} \subset \mathcal{A}_{\alpha,N}^{(n)}$ for all $N, n \in \mathbb{N}$. Notice that we may also write $\mathcal{A}_{\alpha,N}^{(n+1)}$ in the following way:

$$\mathcal{A}_{\alpha,N}^{(n+1)} = \bigcup_{\substack{\ell_1,\ldots,\ell_{n+1} \\ \ell_i \leq N, \, 1 \leq i \leq n+1}} C_\alpha(\ell_1, \ldots, \ell_{n+1}) = \bigcup_{\substack{\ell_1,\ldots,\ell_n \\ \ell_i \leq N, \, 1 \leq i \leq n}} \bigcup_{k \leq N} C_\alpha(\ell_1, \ldots, \ell_n, k).$$

Thus, for all $n \in \mathbb{N}$, we have that

$$\lambda\left(\mathcal{A}_{\alpha,N}^{(n+1)}\right) = \sum_{k=1}^{N} a_k \lambda\left(\mathcal{A}_{\alpha,N}^{(n)}\right).$$

Hence, on applying this argument $n - 1$ more times, it follows that

$$\lambda\left(\mathcal{A}_{\alpha,N}^{(n+1)}\right) = \left(\sum_{k=1}^{N} a_k\right)^n \lambda\left(\mathcal{A}_{\alpha,N}^{(1)}\right).$$

Since the last term above is simply a constant and since $0 < \sum_{k=1}^{N} a_k < 1$, this shows that $\lambda(\mathcal{A}_{\alpha,N}) = 0$, for any $N \in \mathbb{N}$. Finally, we have that

$$\lambda\left(\bigcup_{N=1}^{\infty} \mathcal{A}_{\alpha,N}\right) \leq \sum_{N=1}^{\infty} \lambda(\mathcal{A}_{\alpha,N}) = 0.$$

This finishes the proof of the lemma. $\qquad\square$

Corollary 1.4.22. *Let $\mathcal{W}_\alpha$ be the set defined by*

$$\mathcal{W}_\alpha := \{x = [\ell_1, \ell_2, \ldots]_\alpha \in \mathbb{I}_\alpha : \limsup_{n \to \infty} \ell_n = \infty\}.$$

Then, $\mathcal{W}_\alpha$ is of full Lebesgue measure.

Proof. Notice that the complement of $\mathcal{W}_\alpha$ is the set of all those α-irrational numbers with bounded α-Lüroth elements. The corollary then follows directly from Lemma 1.4.21. $\qquad\square$

Although the sets $\mathcal{A}_{\alpha,N}$ defined in the proof of Lemma 1.4.21 have Lebesgue measure zero for every $N \in \mathbb{N}$, we can still distinguish between their sizes by calculating their Hausdorff dimension. Luckily, this is very easy to do, as the next lemma demonstrates. The reader unfamiliar with the Hausdorff dimension of a set can either refer, for instance, to the book by Falconer [Fal14], or can safely ignore the next two results, as they will not be needed for anything that follows.

Lemma 1.4.23. *For each $N \in \mathbb{N}$, we have*

$$\dim_H\left(\mathcal{A}_{\alpha,N}\right) = s, \text{ where } s \text{ is given by } \sum_{i=1}^{N} a_i^s = 1.$$

Proof. All that is required to prove this statement is to notice that for each $N \in \mathbb{N}$ the set $\mathcal{A}_{\alpha,N}$ is an invariant set for a finite iterated function system $\{L_{\alpha,1}, \ldots, L_{\alpha,N}\}$, where $L_{\alpha,n}$ denotes the n-th inverse branch of the map L_α. Recall that these are given by $L_{\alpha,n}(x) := t_n - a_n x$. That $\mathcal{A}_{\alpha,N}$ is an invariant set for this system means that

$$\mathcal{A}_{\alpha,N} = \bigcup_{i=1}^{N} L_{\alpha,i}\left(\mathcal{A}_{\alpha,N}\right).$$

Then, since these inverse branches are contracting similarities, that is, they satisfy the equality $|L_{\alpha,i}(x) - L_{\alpha,i}(y)| = a_i|x - y|$ for all $x, y \in [0, 1]$, we have that the dimension of $\mathcal{A}_{\alpha,N}$ can be deduced directly from an application of Hutchinson's Formula (see [Fal14], Theorem 9.3). $\qquad\square$

This observation can be used to calculate the Hausdorff dimension of the set $\mathcal{B}_\alpha$, as follows.

Proposition 1.4.24. *Let α be an arbitrary partition of $[0, 1]$. Then,*

$$\dim_H(\mathcal{B}_\alpha) = 1.$$

Proof. Since $\mathcal{B}_\alpha := \bigcup_{N \in \mathbb{N}} \mathcal{A}_N$, we have that

$$\dim_H(\mathcal{B}_\alpha) = \sup \left\{ \dim_H(\mathcal{A}_N) : N \in \mathbb{N} \right\}.$$

Then, by Lemma 1.4.23, $\dim_H(\mathcal{A}_{\alpha,N}) = s$, where s is given by $\sum_{i=1}^N a_i^s = 1$ and $\dim_H(\mathcal{A}_{\alpha,N+1}) = t$, where t is given by $\sum_{i=1}^{N+1} a_i^t = 1$. Therefore, $a_1^t + \cdots + a_N^t < 1$ and so $s < t$. In other words,

$$\dim_H(\mathcal{A}_{\alpha,N}) < \dim_H(\mathcal{A}_{\alpha,N+1}).$$

Furthermore, as $\sum_{i=1}^\infty a_i = 1$, it follows that $\dim_H(\mathcal{B}_\alpha) = 1$. $\square$

Note that similarly, the Hausdorff dimension of the set of badly approximable numbers (for the continued fraction expansion) is also known to be equal to 1. However, the proof is much more involved, so we simply refer to [Jar29]. Let us now consider the result analogous to Theorem 1.2.19.

Theorem 1.4.25.

(a) *Let $\varphi : \mathbb{N} \to \mathbb{N}$ be a function such that the series $\sum_{n=1}^\infty t_{\varphi(n)}$ diverges. Where the set $\mathcal{B}_{\alpha,\varphi}$ is defined by*

$$\mathcal{B}_{\alpha,\varphi} := \{x = [\ell_1, \ell_2, \ldots]_\alpha \in \mathbb{I}_\alpha : \ell_k < \varphi(k) \text{ for all } k \in \mathbb{N}\},$$

we have that

$$\lambda(\mathcal{B}_{\alpha,\varphi}) = 0.$$

(b) *Let $\varphi : \mathbb{N} \to \mathbb{N}$ be a function such that the series $\sum_{n=1}^\infty t_{\varphi(n)}$ converges. Where the set $\mathcal{W}_{\alpha,\varphi}$ is defined to be*

$$\mathcal{W}_{\alpha,\varphi} := \{x = [\ell_1, \ell_2, \ldots]_\alpha \in \mathbb{I}_\alpha : \ell_k > \varphi(k) \text{ infinitely often}\},$$

we have that

$$\lambda(\mathcal{W}_{\alpha,\varphi}) = 0.$$

Proof. For the proof of part (a), we proceed similarly to the proof of Lemma 1.4.21. Define the sets $\mathcal{B}_{\alpha,\varphi}^n$ by setting

$$\mathcal{B}_{\alpha,\varphi}^{(n)} := \{x = [\ell_1, \ell_2, \ldots]_\alpha \in \mathbb{I}_\alpha : \ell_k < \varphi(k) \text{ for all } 1 \le k \le n\}.$$

Then, $\mathcal{B}_{\alpha,\varphi}^{(n+1)} \subset \mathcal{B}_{\alpha,\varphi}^{(n)}$ and $\mathcal{B}_{\alpha,\varphi} \subset \mathcal{B}_{\alpha,\varphi}^{(n)}$ for all $n \in \mathbb{N}$. So, in order to prove that $\lambda(\mathcal{B}_{\alpha,\varphi}) = 0$, it suffices to show that

$$\lim_{n\to\infty} \lambda\left(\mathcal{B}_{\alpha,\varphi}^{(n)}\right) = 0.$$

To that end, notice that for arbitrary $(\ell_1, \ldots, \ell_n) \in \mathbb{N}^n$, we can write

$$\lambda\left(\bigcup_{1 \le k < \varphi(n+1)} C_\alpha(\ell_1, \ldots, \ell_n, k)\right) = \left(\sum_{k=1}^{\varphi(n+1)-1} a_k\right) \lambda(C_\alpha(\ell_1, \ldots, \ell_n))$$

$$= \left(1 - t_{\varphi(n+1)}\right) \lambda(C_\alpha(\ell_1, \ldots, \ell_n)).$$

Thus,

$$\lambda\left(\mathcal{B}_{\alpha,\varphi}^{(n+1)}\right) = \left(1 - t_{\varphi(n+1)}\right) \lambda\left(\mathcal{B}_{\alpha,\varphi}^{(n)}\right) = \cdots = \prod_{k=1}^{n} \left(1 - t_{\varphi(k+1)}\right) \lambda\left(\mathcal{B}_{\alpha,\varphi}^{(1)}\right).$$

Now, since $1 - x \le e^{-x}$ for all $0 < x < 1$, we then have that

$$\lambda\left(\mathcal{B}_{\alpha,\varphi}^{(n+1)}\right) \le e^{-\sum_{k=1}^{n} t_{\varphi(k+1)}} \lambda\left(\mathcal{B}_{\alpha,\varphi}^{(1)}\right).$$

Consequently, as the series $\sum_{k=1}^{n} t_{\varphi(k+1)}$ can be made arbitrarily large as n increases, we have that $\lim_{n\to\infty} \lambda\left(\mathcal{B}_{\alpha,\varphi}^{(n)}\right) = 0$.

Concerning the proof of part (b), we will again use the Borel–Cantelli Lemma. Notice that

$$\mathcal{W}_{\alpha,\varphi} = \limsup_{n\to\infty} \mathcal{W}_{\alpha,\varphi}^{(n)}, \quad \text{where } \mathcal{W}_{\alpha,\varphi}^{(n)} := \{x \in \mathbb{I}_\alpha : \ell_n \ge \varphi(n)\}.$$

Thus, to finish the proof, it is enough to show that

$$\sum_{n=1}^{\infty} \lambda\left(\mathcal{W}_{\alpha,\varphi}^{(n)}\right) < \infty.$$

Indeed,

$$\lambda\left(\mathcal{W}_{\alpha,\varphi}^{(n)}\right) = \lambda\left(\bigcup_{(\ell_1,\ldots,\ell_{n-1})\in\mathbb{N}^n} \bigcup_{k:k\ge\varphi(n)} C_\alpha(\ell_1, \ldots, \ell_{n-1}, k)\right)$$

$$= \sum_{(\ell_1,\ldots,\ell_{n-1})\in\mathbb{N}^n} \sum_{k:k\ge\varphi(n)} \lambda(C_\alpha(\ell_1, \ldots, \ell_{n-1}, k))$$

$$= \sum_{(\ell_1,\ldots,\ell_{n-1})\in\mathbb{N}^n} \sum_{k:k\ge\varphi(n)} a_{\ell_1} \ldots a_{\ell_{n-1}} a_k = t_{\varphi(n)}.$$

By assumption, the series $\sum_{n=1}^{\infty} t_{\varphi(n)}$ converges and so, therefore, does the series $\sum_{n=1}^{\infty} \lambda \left(\mathcal{W}_{\alpha,\varphi}^{(n)} \right)$. This finishes the proof. □

Remark 1.4.26. Notice that if the partition α is of finite type, that is, if α is such that $\sum_{n=1}^{\infty} t_n$ converges, then we have that $\lambda(\mathcal{B}_{\alpha,\varphi}) = 1$ for any arbitrary increasing function $\varphi : \mathbb{N} \to \mathbb{N}$.

1.5 Notes and historical remarks

1.5.1 The Farey sequence

The Farey map is named for John Farey (1766–1826), who was not a mathematician, but a geologist. Farey's one contribution to Mathematics was the article *On a curious property of vulgar fractions* [Far16], in which he defines Farey sequences in the following way. For each $n \in \mathbb{N}$, list all the rationals between 0 and 1 which, when expressed in their lowest terms, have denominator at most equal to n. Denoting the *n-th Farey sequence* by $\mathcal{F}_n$, the first few are given by

$$\mathcal{F}_1 := \left\{ \frac{0}{1}, \frac{1}{1} \right\}, \quad \mathcal{F}_2 := \left\{ \frac{0}{1}, \frac{1}{2}, \frac{1}{1} \right\}, \quad \mathcal{F}_3 := \left\{ \frac{0}{1}, \frac{1}{3}, \frac{1}{2}, \frac{2}{3}, \frac{1}{1} \right\},$$

$$\mathcal{F}_4 := \left\{ \frac{0}{1}, \frac{1}{4}, \frac{1}{3}, \frac{1}{2}, \frac{2}{3}, \frac{3}{4}, \frac{1}{1} \right\}, \quad \mathcal{F}_5 := \left\{ \frac{0}{1}, \frac{1}{5}, \frac{1}{4}, \frac{1}{3}, \frac{2}{5}, \frac{1}{2}, \frac{3}{5}, \frac{2}{3}, \frac{3}{4}, \frac{1}{1} \right\}, \dots$$

The curious property of Farey's title is that each member of the sequence is equal to the *mediant* of its two neighbours. Recall that the mediant of two rational numbers a/b and a'/b' is by definition the rational number $(a + a')/(b + b')$. Farey did not himself provide a proof of his discovered property[2] and he was doubtless not the first to notice it. Cauchy supplied the necessary proof in the same year that Farey's article appeared.

We have already seen that if we iterate the point $1/2$ under the two inverse branches of the Farey map, each time one of the Farey fractions turns up. However, as we have already pointed out, strictly speaking it is not the Farey sequence which appears in this manner, but rather the Stern–Brocot sequence. The Stern–Brocot sequence was independently discovered by the German number-theorist Moritz Stern [Ste58] and the French clockmaker Achille Brocot [Bro61]. (Brocot used the sequences to design systems of gears. For information on these sorts of applications, see Chapter IV of Rockett and Szűsz [RS92].) For this reason, it would perhaps be more reasonable to refer to the Farey map as the Stern–Brocot map. However, we choose to stick with convention on this point.

---

[2] That Farey did not give a proof of his curious property was pointed out by Hardy [HW08], with the rather unfriendly comment that Farey was "at the best an indifferent mathematician".

1.5.2 The classical Lüroth series

In the paper [Lür83], J. Lüroth introduces a series representation of real numbers from the unit interval. His starting point is the observation that for every real number x in the interval $(0, 1)$, either $x = 1/\ell$, for some positive integer $\ell \geq 2$, or, $1/x$ lies between two successive positive integers ℓ_1 and $\ell_1 + 1$ and so

$$x = \frac{1}{\ell_1 + 1} + \widehat{x},$$

where, since $x < 1/\ell_1$, we have that $0 < \widehat{x} < 1/(\ell_1(\ell_1 + 1))$. Now, defining $x_1 := \widehat{x}(\ell_1 + 1)\ell_1$ supplies the equation

$$x = \frac{1}{\ell_1 + 1} + \frac{x_1}{\ell_1(\ell_1 + 1)}.$$

Note that since $0 < \widehat{x} < 1/(\ell_1(\ell_1 + 1))$, we also obtain the inequality $0 < x_1 < 1$. Therefore, the same argument holds for x_1 as for the original point x, which leads to the equation

$$x = \frac{1}{\ell_1 + 1} + \frac{1}{\ell_1(\ell_1 + 1)(\ell_2 + 1)} + \frac{x_2}{\ell_1(\ell_1 + 1)\ell_2(\ell_2 + 1)}.$$

Clearly, this process either continues until such a time as one of the x_i is equal to the reciprocal of a positive integer that is at least equal to 2, or continues indefinitely. For the special case that $x = 1$, we notice that $1 = 1/2 + 1/4 + 1/8 + \ldots$. In each case, this gives the series expansion now called the *Lüroth expansion* of a real number in $[0, 1]$.

Each finite expansion of the form above represents a rational number. Suppose now that $x \in [0, 1]$ has an infinite Lüroth expansion. Since each ℓ_k is at least equal to 1, for the k-th term in the Lüroth expansion of x we have that

$$\frac{1}{\ell_1(\ell_1 + 1)\ldots\ell_{k-1}(\ell_{k-1} + 1)(\ell_k + 1)} \leq \frac{1}{2^k}.$$

Thus, it makes sense to write

$$x = \sum_{n=1}^{\infty} \left(\ell_n \prod_{k=1}^{n} (\ell_k(\ell_k + 1))^{-1} \right).$$

We will use Lüroth's original notation and write $x = S(\ell_1, \ell_2, \ldots)$ for this sum. For instance, we have that $1 = S(1, 1, 1, \ldots)$. The next observation in [Lür83] is that if $x \in [0, 1]$ has a finite Lüroth expansion, that is, if $x = S(\ell_1, \ell_2, \ldots, \ell_k)$ for some $k \in \mathbb{N}$, then

$$x = S(\ell_1, \ell_2, \ldots, (\ell_k + 1), 1, 1, 1, \ldots).$$

This is straightforward to check. It follows that every number x in $(0, 1]$ has an infinite Lüroth expansion. In fact, this infinite representation is unique. As already mentioned, each finite Lüroth expansion represents a rational number, but it is easy to see, using only the sum of a geometric series, that each (eventually) periodic infinite Lüroth expansion is also a rational number. Of course, each finite Lüroth expansion can also be written as an eventually periodic expansion; in this case the periodic part consists of infinitely many ones. The proof of these statements are also given in [Lür83].

It seems probable that Lüroth was thinking of a generalisation of the decimal expansion of a real number when he introduced his infinite series expansion. He states that the given expansion has many similarities with the representation through infinite decimal expansions and asks whether or not it is possible to characterise the numbers which have a finite Lüroth expansion in any other way, that is, as in the way that rational numbers with finite decimal representations are exactly those with denominators equal to $2^n 5^m$ for some positive integers n and m. As of the present moment, we are unaware of any answer to this question.

The Lüroth expansion can also be generated by a dynamical system, $L : [0, 1] \to [0, 1]$. The map L is referred to as the *Lüroth map* and it is defined by

$$
L(x) := \begin{cases} n(n+1)x - n & \text{for } x \in \left[\dfrac{1}{n+1}, \dfrac{1}{n}\right), \ n \geq 2; \\ 2x - 1 & \text{for } x \in \left[\dfrac{1}{2}, 1\right]; \\ 0 & \text{for } x = 0. \end{cases}
$$

The graph of the map L is shown in Fig. 1.10 below.

The Lüroth expansion of a real number in $[0, 1]$ is generated by the Lüroth map in precisely the same way as in all the other examples given above. It can be seen from the graph of the map L that it is (basically) nothing other than the map L_{α_H} with

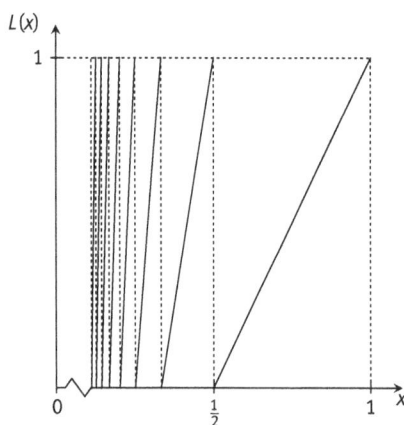

Fig. 1.10. The Lüroth map, $L : [0, 1] \to [0, 1]$.

all positive slopes instead of all negative slopes. The map L_{α_H} was first described by S. Kalpazidou, A. Knopfmacher and J. Knopfmacher [KKK91] in the early 1990s; they called it the *alternating Lüroth map* and established some of its basic properties. The Lüroth map and, to a lesser extent, the alternating Lüroth map have been studied by several authors. In addition to those already cited above, these works include [BBDK96], [DK96], [Gal72], [Gan01], [Šal68], [SW07] and in particular [DK02].

1.6 Exercises

Exercise 1.6.1 (Dirichlet's Approximation Theorem). Fix $x \in \mathbb{R}$. Prove that for every $N \in \mathbb{N}$ there exists $p, q \in \mathbb{Z}$ with $1 \le q \le N$ such that

$$|xq - p| \le \frac{1}{N}$$

and deduce that for infinitely many co-prime integers p and q we have that

$$\left| x - \frac{p}{q} \right| \le \frac{1}{q^2}.$$

Hint: Apply the *Pigeonhole Principle* to the 'pigeons' $kx - \lfloor kx \rfloor$, for $k = 0, \ldots, N$ and the 'holes' $[\ell/N, (\ell+1)/N)$, $\ell = 0, \ldots, N-1$.

Exercise 1.6.2. Recall the following good decimal approximation of π:

$$\pi \approx 3.141592653589793\ldots$$

(i) Find the first four elements in the continued fraction expanion of π.
(ii) Determine the first four convergents of π.

Exercise 1.6.3. Let $x, y \in \mathbb{R}$, $A, B, C \in \mathbb{Z}$, and $a, b, c, d \in \mathbb{N}$ such that $cy + d \ne 0$. Show that if

$$x = \frac{ay + b}{cy + d} \quad \text{and} \quad Ay^2 + By + C = 0,$$

then x satisfies

$$Dx^2 + Ex + F = 0, \quad \text{for some } D, E, F \in \mathbb{Z}.$$

Determine D, E and F in terms of A, B, C, a, b, c and d.

Exercise 1.6.4. Let (f_n) denote the Fibonacci sequence, that is, $f_0 := 1$, $f_1 := 1$ and $f_{n+2} := f_{n+1} + f_n$. Show that for the generating function we have

$$\sum_{k=0}^{\infty} f_k z^k = \frac{z}{1 - z - z^2}$$

and deduce that

$$f_n := \frac{1}{\sqrt{5}} \left(\gamma^n - (-\gamma)^{-n} \right),$$

where $\gamma := (1 + \sqrt{5})/2$ denotes the golden mean.

Exercise 1.6.5. Taking inspiration from the proof of Hurwitz's Theorem II, prove that for $x := \sqrt{2} - 1 = [2, 2, 2, \ldots]$, we have $\nu(x) = 1/\sqrt{8}$.

Exercise 1.6.6. Prove that if x and y are two equivalent irrational numbers (in the sense of Definition 1.1.13 (c)), then $\nu(x) = \nu(y)$.

Exercise 1.6.7. Let x and y be two equivalent irrational numbers. Show that there exist integers a, b, c and d such that

$$x = \frac{ay + b}{cy + d},$$

with $ad - bc = \pm 1$.

Exercise 1.6.8. Show that for every $n \in \mathbb{N}$, the continued fraction expansion of $\sqrt{n^2 + 1} - n$ is given by the periodic expansion

$$\sqrt{n^2 + 1} - n = [\overline{2n}] = \cfrac{1}{2n + \cfrac{1}{2n + \ldots}}.$$

Exercise 1.6.9. Let $e = \sum_{k=0}^{\infty} 1/k! = 2.71828\ldots$ be the base of the natural logarithm, and note that is known that the continued fraction expansion of e is given by

$$e = a_0 + [a_1, a_2, \ldots] = 2 + [1, 2, 1, 1, 4, 1, 1, 6, 1, 1, 8, \ldots],$$

where $a_0 := 2$, $a_1 := 1$, and for $n \geq 1$ we have that $a_{3n-1} := 2n$ and $a_{3n} = a_{3n+1} := 1$. Show that the following statements are true for every integer $n \geq 1$.
 (i) For exactly one element $i \in \{n, n+1, n+2\}$, we have that

$$\left| e - \frac{p_i}{q_i} \right| < \frac{3}{2(i+2)q_i^2}.$$

(ii) For exactly two elements $i \in \{n, n+1, n+2\}$, we have that

$$\left| e - \frac{p_i}{q_i} \right| > \frac{1}{3q_i^2}.$$

Exercise 1.6.10. Show that the function $d : E^{\infty} \times E^{\infty} \to [0, 1]$ defined in Definition 1.2.10 is really a metric.

Exercise 1.6.11. Show that the space $\mathbb{N}^{\mathbb{N}}$ equipped with the metric d from Definition 1.2.10 is a complete metric space which is not locally compact.

Exercise 1.6.12. Show that the map $\psi : \mathbb{I} \to \mathbb{N}^{\mathbb{N}}$ defined by

$$\psi([x_1, x_2, x_3, \ldots]) := (x_1, x_2, x_3, \ldots)$$

is a topological conjugacy map between the Gauss system and the full shift map on the shift space $\mathbb{N}^{\mathbb{N}}$.

Exercise 1.6.13. Show that the tent map and the map $L : [0, 1] \to [0, 1]$, $x \mapsto 4x(1 - x)$, are conjugated via

$$\psi : [0, 1] \to [0, 1], x \mapsto \frac{1}{2} - \frac{1}{2}\cos(\pi x).$$

Exercise 1.6.14. Prove Cantor's Intersection Theorem: If $(S_n)_{n \in \mathbb{N}}$ is a decreasing sequence (so $S_{n+1} \subseteq S_n$) of non-empty compact sets in $\mathbb{R}$ (more generally, a complete metric space), with diameters shrinking to zero, then the intersection $S := \bigcap_{n \in \mathbb{N}} S_n$ is a singleton.

Hint: Choose a sequence of points $x_n \in S_n$. Show that $(x_n)_{n \in \mathbb{N}}$ is a Cauchy sequence and therefore converges, to x say. Show that this x lies in each set S_n. Finally, show that if $x, y \in S$, then $x = y$.

Exercise 1.6.15. Show that Minkowski's question-mark function Q maps the set of rational numbers onto the dyadic rationals and maps the quadratic surds onto the set of non-dyadic rationals and show that their order is preserved.

Exercise 1.6.16. Re-derive the formula given by Denjoy for the function Q from the definition given in terms of mediants.

Hint: Find inspiration in the proof of Proposition 1.3.11.

Exercise 1.6.17. Prove Proposition 1.4.12.

Exercise 1.6.18. Provide the missing details in the proof that the map L_α is κ_+-Hölder continuous.

Exercise 1.6.19. Supply the missing details in the proof that L_α is κ_--sub-Hölder continuous.

Exercise 1.6.20. Find an operator K on the set $\mathcal{S}$ of continuous, surjective, strictly increasing functions defined on the unit interval such that $K^n(f)$ converges uniformly to Minkowski's question-mark function Q for any $f \in \mathcal{S}$.

2 Basic ergodic theory

In this chapter, we aim to investigate the basic measure- and ergodic-theoretic properties of all our main examples. Along the way, we will recall some standard definitions and theorems that will be of use to us when we come to consider these particular examples. We will assume that the reader is somewhat familiar with measures and basic measure theory, but any definitions or theorems that we do not mention explicitly can be found in, for instance, either [Coh80] or [Rud87].

2.1 Invariant measures

From this point on, we are interested in *(measure-theoretic) dynamical systems*, in other words, systems $(X, \mathcal{B}, \mu, T)$ where $(X, \mathcal{B}, \mu)$ is a σ-finite measure space and $T : X \to X$ is a measurable transformation. The most fundamental objects of interest in ergodic theory are measures which remain invariant under the dynamics of the system. These are the subject of our first definition.

Definition 2.1.1. Let $(X, \mathcal{B}, \mu, T)$ be a dynamical system. The measure μ is called *T-invariant* if we have

$$\mu \circ T^{-1}(A) := \mu(T^{-1}(A)) = \mu(A) \text{ for all } A \in \mathcal{B}.$$

We also say that $T : X \to X$ is a *measure-preserving transformation* and call the system $(X, \mathcal{B}, \mu, T)$ a *measure-preserving system*.

Remark 2.1.2. For every T-invariant measure μ and any set $A \in \mathcal{B}$ of finite μ-measure, we have

$$\mu(T^{-1}(A) \setminus A) = \mu(T^{-1}(A)) - \mu(T^{-1}(A) \cap A)$$
$$= \mu(A) - \mu(T^{-1}(A) \cap A) = \mu(A \setminus T^{-1}(A)).$$

In practice, it can be difficult to check that a map preserves a given measure using only this definition, as it is often the case that no specific information is known about a general measurable set. However, it is enough to have knowledge of a particular class of sets that generates the σ-algebra of measurable sets, as we now show. (Recall that the σ-algebra generated by a collection $\mathcal{C}$ of subsets of X is the smallest, in the sense of inclusion, σ-algebra that contains all the sets in $\mathcal{C}$.)

Lemma 2.1.3. *Let $(X, \mathcal{B}, \mu)$ be a measure space and let $T : X \to X$ be a measurable function. Suppose that S is a collection of subsets of X closed under taking intersections with $\sigma(S) = \mathcal{B}$ and such that μ is σ-finite on S. Then, if $\mu \circ T^{-1}(B) = \mu(B)$ for every set $B \in S$, we have that the map T preserves the measure μ.*

Proof. Since T is measurable, we have that $\mu \circ T^{-1}$ defines a measure on $\mathcal{B}$. Moreover, this measure coincides with μ on S. The facts that S is closed under taking intersections and generates $\mathcal{B}$ and that μ is σ-finite on S guarantee that the measures μ and $\mu \circ T^{-1}$ coincide on $\mathcal{B}$. $\qquad\square$

Example 2.1.4. The collection of all subintervals of $[0, 1]$ together with the empty set generates the Borel σ-algebra on $[0, 1]$ and is closed under taking intersections.

To motivate the definition of T-invariance from the stochastic point of view let us consider a probability space $(X, \mathcal{B}, \mu)$ together with a measurable transformation $T : X \to X$. Then the stochastic process given by $(g \circ T^k)_{k \in \mathbb{N}_0}$ defined on $(X, \mathcal{B}, \mu)$ is stationary for every integrable function $g : X \to \mathbb{R}$ if and only if μ is T-invariant. Stationarity follows from T-invariance by noting that for every Borel set B and every $k \in \mathbb{N}$ we have

$$\mu(\{x \in X : g \circ T^k(x) \in A\}) = \mu(T^{-k}\{x \in X : g(x) \in A\}) = \mu(\{x \in X : g(x) \in A\}),$$

if μ is T-invariant. Stationarity implies T-invariance by taking $g := \mathbb{1}_A$ for a measurable set A and $k = 1$ and observing that

$$\mu(T^{-1}(A)) = \mathbb{E}_\mu(\mathbb{1}_A \circ T) = \mathbb{E}_\mu(\mathbb{1}_A) = \mu(A).$$

Example 2.1.5. For the tent map $T : [0, 1] \to [0, 1]$ it is easily seen that for any interval $(a, b) \in [0, 1]$ we have that

$$\lambda(T^{-1}(a, b)) = \lambda\left(\left(\frac{a}{2}, \frac{b}{2}\right) \cup \left(\frac{2-b}{2}, \frac{2-a}{2}\right)\right)$$
$$= \frac{1}{2}(b - a) + \frac{1}{2}(b - a) = b - a = \lambda((a, b)).$$

Thus, it follows from Lemma 2.1.3 that the tent map preserves the Lebesgue measure.

The following definitions will be used throughout the book.

Definition 2.1.6. Let $(X, \mathcal{B}, \mu)$ be a measure space and let $\mathcal{M}(\mathcal{B})$ denote the set of all real-valued $\mathcal{B}$-measurable functions. We now define an equivalence relation by identifying two elements $f, g \in \mathcal{M}(\mathcal{B})$ if they satisfy

$$f \sim g : \iff \mu(\{f \neq g\}) = 0. \tag{2.1}$$

Then we define

$$M(\mathcal{B}) := \mathcal{M}(\mathcal{B})/\sim.$$

Further, define

$$L_1(\mu) := \left\{ f : X \to \mathbb{R} : f \text{ measurable and } \int |f| \, d\mu < \infty \right\} / \sim$$

and

$$L_\infty(\mu) := \{ f : X \to \mathbb{R} : f \text{ measurable and } \operatorname{ess\,sup}|f| < \infty \} / \sim .$$

If it is clear from the context, we will not distinguish between measurable functions and their equivalence classes given by (2.1). For any subset $\mathcal{F}$ of $M(\mathcal{B})$, we let $\mathcal{F}^+$ denote the subset of non-negative elements from $\mathcal{F}$.

For the case of a finite measure space, the following proposition gives a useful characterisation of T-invariant measures in terms of the space of $L_1(\mu)$ functions.

Proposition 2.1.7. *Let $T : (X, \mathcal{B}, \mu) \to (X, \mathcal{B}, \mu)$ be a measurable transformation on a probability space $(X, \mathcal{B}, \mu)$. Then μ is T-invariant if and only if*

$$\int f \, d\mu = \int f \circ T \, d\mu, \quad \text{for all } f \in L_1(\mu). \tag{2.2}$$

Proof. Suppose first that (2.2) holds. In that case, for any measurable set B we can let $f = \mathbb{1}_B$ to obtain that

$$\mu(B) = \int \mathbb{1}_B \, d\mu = \int \mathbb{1}_B \circ T \, d\mu = \mu(T^{-1}(B)).$$

On the other hand, if T preserves the measure μ, then (2.2) holds for all characteristic functions and therefore it also holds for all simple functions. If $f \in L_1(\mu)$, we can find a sequence $(f_n)_{n \geq 1}$ of simple functions increasing to f; moreover, the sequence $(f_n \circ T)_{n \geq 1}$ is a sequence of simple functions increasing to $f \circ T$. We can then apply the Monotone Convergence Theorem to deduce that

$$\int f \, d\mu = \lim_{n \to \infty} \int f_n \, d\mu = \lim_{n \to \infty} \int f_n \circ T \, d\mu = \int f \circ T \, d\mu.$$

Since $f \in L_1(\mu)$ was arbitrary, the proof is finished. $\qquad\square$

The proof of Proposition 2.1.7 actually shows rather more, namely, that it is enough to check that the condition holds for all bounded measurable functions.

2.1.1 Invariant measures for the Gauss and α-Lüroth system

Our aim now is to describe the invariant measures for our main examples. Let us begin with the α-Lüroth systems.

Proposition 2.1.8. *The Lebesgue measure λ is L_α-invariant.*

Proof. Recall from Section 1.4.1 that the inverse branches $L_{\alpha,n} : [0,1) \to A_n$ of L_α are given by $L_{\alpha,n}(x) := t_n - a_n x$, for all $n \in \mathbb{N}$. In order to show that λ is L_α-invariant, by Lemma 2.1.3 it is enough to show that for every subinterval I contained in $[0,1]$, we have that

$$\lambda(I) = \lambda(L_\alpha^{-1}(I)).$$

In fact, since the Lebesgue measure is non-atomic, it suffices to let $I := [a,b]$ for some $0 \le a < b \le 1$. A straightforward calculation shows that

$$\lambda(L_\alpha^{-1}[a,b]) = \lambda\left(\bigcup_{n=1}^{\infty} L_{\alpha,n}([a,b])\right) = \sum_{n=1}^{\infty} \lambda(L_{\alpha,n}([a,b]))$$

$$= \sum_{n=1}^{\infty} |(t_n - a_n a) - (t_n - a_n b)| = \sum_{n=1}^{\infty} a_n(b-a)$$

$$= b - a = \lambda([a,b]). \qquad \square$$

As you will show in Exercise 2.6.1, the Gauss map is not preserved by the Lebesgue measure. However, Gauss observed in 1845 that the map G does preserve a Lebesgue-absolutely continuous measure, which we now define.

Definition 2.1.9. The *Gauss measure m_G* is given, for all Borel measurable sets $A \subseteq [0,1]$, by

$$m_G(A) := \frac{1}{\log 2} \int_A \frac{1}{1+x} \, d\lambda(x).$$

Proposition 2.1.10. *The Gauss map G preserves the Gauss measure m_G.*

Proof. It suffices to show that $m_G([0,b]) = m_G \circ G^{-1}([0,b])$, for any $0 \le b \le 1$. Recalling that G_n refers to the n-th inverse branch of G, which is given by $G_n(x) := 1/(x+n)$, we have that

$$G^{-1}([0,b]) = \bigcup_{n=1}^{\infty} G_n([0,b]) = \bigcup_{n=1}^{\infty} \left[\frac{1}{n+b}, \frac{1}{n}\right].$$

Thus, first observing that

$$\frac{1 + \dfrac{1}{n}}{1 + \dfrac{1}{n+b}} = \frac{\dfrac{n+1}{n} \cdot \dfrac{n+b}{n+1}}{\dfrac{n+b+1}{n+b} \cdot \dfrac{n+b}{n+1}} = \frac{1 + \dfrac{b}{n}}{1 + \dfrac{b}{n+1}},$$

the following calculation finishes the proof:

$$m_G(G^{-1}([0, b])) = \frac{1}{\log 2} \sum_{n=1}^{\infty} \int_{1/(n+b)}^{1/n} \frac{1}{1+x} \, dx$$

$$= \frac{1}{\log 2} \sum_{n=1}^{\infty} \left(\log \left(1 + \frac{1}{n} \right) - \log \left(1 + \frac{1}{n+b} \right) \right)$$

$$= \frac{1}{\log 2} \sum_{n=1}^{\infty} \left(\log \left(1 + \frac{b}{n} \right) - \log \left(1 + \frac{b}{n+1} \right) \right)$$

$$= \frac{1}{\log 2} \sum_{n=1}^{\infty} \int_{b/(n+1)}^{b/n} \frac{1}{1+x} \, dx$$

$$= \sum_{n=1}^{\infty} m_G \left(\left[\frac{b}{n+1}, \frac{b}{n} \right] \right)$$

$$= m_G([0, b]). \qquad \square$$

Remark 2.1.11. In order to make a guess at how Gauss arrived at his invariant measure, we make the simple arithmetic observation that

$$\sum_{n=1}^{\infty} \frac{1}{(n+x)(n+1+x)} = \sum_{n=1}^{\infty} \left(\frac{1}{n+x} - \frac{1}{n+1+x} \right) = \frac{1}{1+x},$$

which is equivalent to

$$\sum_{n=1}^{\infty} \frac{1}{(n+x)^2} \frac{1}{1 + \frac{1}{n+x}} = \frac{1}{1+x}.$$

This infinite sum can be expressed in terms of the Gauss map G as follows, with $h_G(x) := 1/((1+x)\log 2)$, and, as usual, with G_n referring to the n-th inverse branch of G,

$$\sum_{n=1}^{\infty} |G_n'(x)| \, h_G(G_n(x)) = h_G(x).$$

The significance of this formula will be apparent shortly, when we come to describe the transfer operator.

Finally, we finish this section with the observation that the Gauss measure belongs to the same measure class as the Lebesgue measure.

Proposition 2.1.12. *For any Borel set $B \subseteq [0, 1]$, we have that*

$$\frac{\lambda(B)}{2 \log 2} \leq m_G(B) \leq \frac{\lambda(B)}{\log 2}.$$

Proof. Since in the following x lies between 0 and 1, we have that

$$m_G(B) = \frac{1}{\log 2} \int_B \frac{1}{1+x} \, d\lambda(x) \leq \frac{1}{\log 2} \int_B 1 \, d\lambda(x) = \frac{\lambda(B)}{\log 2},$$

and

$$m_G(B) = \frac{1}{\log 2} \int_B \frac{1}{1+x} \, d\lambda(x) \geq \frac{1}{\log 2} \int_B \frac{1}{2} \, d\lambda(x) = \frac{\lambda(B)}{2 \log 2}. \qquad \square$$

2.2 Recurrence and conservativity

Let us now study a general property of invariant measures by taking a detour to present one of the fundamental results of ergodic theory, namely, Halmos's Recurrence Theorem [Hal56]. This theorem states that for a conservative transformation (which will be defined momentarily) on a σ-finite measure space, almost all points of a given set return infinitely often to that set under iteration. Although it is relatively easy to prove, the importance of this theorem should not be underestimated, as it is really one of the very few completely general theorems in all of ergodic theory. Theorems of this type were initially established for finite systems, but the proofs are essentially the same.

Definition 2.2.1. Let $(X, \mathcal{B}, \mu, T)$ be a measure-theoretic dynamical system. A set $W \in \mathcal{B}$ is called a *wandering set* for T if $\{T^{-n}(W) : n \geq 0\}$ is a collection of pairwise disjoint sets. The collection of all wandering sets will be denoted by $\mathcal{W}_T$

Note that if $W \subseteq X$ is a wandering set for a map $T : X \to X$, then this implies that

$$\sum_{n=0}^{\infty} \mathbb{1}_W \circ T^n \leq 1.$$

Definition 2.2.2. Let $(X, \mathcal{B}, \mu)$ be a measure space and $T : X \to X$ a map. Then T is said to be *conservative* if every wandering set for T is a null set for μ. If $(X, \mathcal{B}, \mu, T)$ is a measure-preserving system and T is conservative, then we will call the system a *conservative measure-preserving system*.

Proposition 2.2.3. *Any measure-preserving system $(X, \mathcal{B}, \mu, T)$ with $\mu(X)$ finite is conservative.*

Proof. Fix $W \in \mathcal{B}$ and assume that $T^{-k}(W)$ defines a disjoint family of sets for $k \in \mathbb{N}_0$. Then

$$\mu(X) \geq \mu\left(\bigcup_{k \in \mathbb{N}} T^{-k}W\right) = \sum_{k \in \mathbb{N}} \mu(T^{-k}W) = \sum_{k \in \mathbb{N}} \mu(W)$$

implies that $\mu(W) = 0$. □

Note that infinite measure spaces are not guaranteed to be conservative. Indeed, the simplest counterexample is a translation of the real line. For instance, consider the map given by $T(x) = x + 1$, which certainly preserves the Lebesgue measure on $\mathbb{R}$, and let $A = (0, 1)$. Clearly, no point of A ever comes back to A under iteration of T, although the Lebesgue measure of A is evidently not equal to zero.

We are now ready to prove the main result of this section. For this we make the following definition.

Definition 2.2.4. A transformation $T : X \to X$ of a measure space $(X, \mathcal{B}, \mu)$ is said to be *non-singular* if for each $B \in \mathcal{B}$ with $\mu(B) = 0$ we have that $\mu(T^{-1}(B)) = 0$. That is, the map T^{-1} preserves sets of μ-measure zero. In this case we also call the dynamical system $(X, \mathcal{B}, \mu, T)$ non-singular.

Remark 2.2.5. A system is sometimes said to be (two-sided) non-singular if for each $B \in \mathcal{B}$ we have $\mu(B) = 0$ if and only if $\mu(T^{-1}(B)) = 0$. We never need this stronger property.

Let us recall that the *symmetric difference* $A \triangle B$ of two sets A and B is defined by $A \triangle B := (A \setminus B) \cup (B \setminus A) = (A \cup B) \setminus (A \cap B)$. Hereafter, we will use the notation "$A = B \mod \mu$", (respectively, "$A \subset B \mod \mu$") to indicate that two sets are equal (respectively, A is contained in B) up to a set of μ-measure zero, i.e., $\mu(A \triangle B) = 0$ (respectively, $\mu(A \setminus B) = 0$).

Theorem 2.2.6 (Halmos's Recurrence Theorem). *Let $(X, \mathcal{B}, \mu, T)$ be a non-singular dynamical system. Then for every set $A \in \mathcal{B}$ we have that $\mu(A \cap W) = 0$ for all wandering sets $W \in \mathcal{B}$ if and only if for all measurable subsets $B \subset A$ the set of points from B returning infinitely often under T to B is equal to $B \mod \mu$, i.e.,*

$$\mu\left(\{x \in B : T^n(x) \in B \text{ for infinitely many } n \geq 1\}\right) = \mu(B).$$

Proof. For $B \subset A$ measurable, let

$$N := \{x \in B : T^n(x) \notin B \text{ for all } n \geq 1\} = B \setminus \bigcup_{n=1}^{\infty} T^{-n}B.$$

Our first claim is that $\mu(N) = 0$. To show this, let $x \in N$. Then for every $n \geq 1$ we have that $T^n(x) \notin B$, and therefore $T^n(x) \notin N$. This shows that $N \cap T^{-n}(N) = \emptyset$, for all $n \geq 1$.

Hence, it follows that for all $i, j \in \mathbb{N}$ such that $j < i$, we have that

$$T^{-j}(N) \cap T^{-i}(N) = T^{-j}\left(N \cap T^{-(i-j)}(N)\right) = T^{-j}(\varnothing) = \varnothing.$$

So, the preimages $\{T^{-n}(N) : n = 0, 1, 2, \ldots\}$ of N under the iterates of T form a pairwise disjoint family of sets, that is to say, N is a wandering subset of A. By our assumption on A and since T is assumed to be non-singular, it follows that $0 = \mu(N) = \mu(T^{-n}N)$, for all $n \in \mathbb{N}$. Since $N = B \setminus \bigcup_{n=1}^{\infty} T^{-n}B$ this shows that for all $n \geq 0$ we have

$$T^{-n}B \subset \bigcup_{k>n} T^{-k}B \quad \mathrm{mod}\, \mu.$$

Consequently,

$$\bigcup_{k>n} T^{-k}B = T^{-n-1}B \cup \bigcup_{k>n+1} T^{-k}B = \bigcup_{k>n+1} T^{-k}B \quad \mathrm{mod}\, \mu.$$

From this we deduce that

$$B \subset \bigcup_{k=0} T^{-k}B = \bigcup_{k>1} T^{-k}B = \cdots = \bigcap_{n \in \mathbb{N}} \bigcup_{k \geq n} T^{-k}B$$

$$= \{x \in X : T^n(x) \in B \text{ for infinitely many } n \geq 1\} \quad \mathrm{mod}\, \mu.$$

For the reverse implication we assume that $\mu(A \cap W) > 0$ for some wandering set $W \in \mathcal{B}$. Then for the set $B := A \cap W$ of positive measure we have $T^{-n}B \cap B = \varnothing$, for all $n \in \mathbb{N}$. □

Note that the recurrence property for B in Halmos's Recurrence Theorem can be stated equivalently as follows

$$\sum_{n=1}^{\infty} \mathbb{1}_B \circ T^n = \infty \quad \mu\text{-a.e. on } B.$$

This leads to the following useful observation.

Lemma 2.2.7. *Let $(X, \mathcal{B}, \mu, T)$ be a measure-preserving system and f, g two measurable functions with $g \in L_1^+(\mu)$ and $f > 0$. Then*

$$\left\{x \in X : \sum_{k=0}^{\infty} g \circ T^k(x) = \infty\right\} \subset \left\{x \in X : \sum_{k=0}^{\infty} f \circ T^k(x) = \infty\right\} \quad \mathrm{mod}\, \mu.$$

Proof. Let us consider the increasing sequence of sets $W_f^N := \left\{\sum_{k=0}^{\infty} f \circ T^k < N\right\}$ which have union $\left\{\sum_{k=0}^{\infty} f \circ T^k < \infty\right\}$. Using the fact that $T^{-k}W_f^N \subset W_f^N$, for every $k \in \mathbb{N}$, and the T-invariance of μ we have for every $n \in \mathbb{N}$

$$\int_{W_f^N} \sum_{k=0}^n g \circ T^k f \, d\mu = \sum_{k=0}^n \int \mathbb{1}_{W_f^N} \cdot g \circ T^{n-k} \cdot f \, d\mu$$

$$= \sum_{k=0}^n \int \mathbb{1}_{W_f^N} \circ T^k \cdot f \circ T^k \cdot g \circ T^n \, d\mu$$

$$= \int g \circ T^n \sum_{k=0}^n \mathbb{1}_{T^{-k} W_f^N} \cdot f \circ T^k \, d\mu$$

$$\leq \int g \circ T^n \cdot \mathbb{1}_{W_f^N} \cdot \sum_{k=0}^n f \circ T^k \, d\mu \leq N \int g \, d\mu < \infty.$$

Since $f > 0$ this shows that μ-a.e. on W_f^N the infinite sum $\sum_{k=0}^\infty g \circ T^k$ is finite. Hence, $\left\{ \sum_{k=0}^\infty f \circ T^k < \infty \right\} = \bigcup_N W_f^N \subset \left\{ \sum_{k=0}^\infty g \circ T^k < \infty \right\}$. Taking complements proves the inclusion. $\qquad\square$

Now Lemma 2.2.7 gives rise to the definition of the *Hopf decomposition* of X with respect to a measure-preserving transformation T.

Definition 2.2.8 (Hopf decomposition). Let $(X, \mathcal{B}, \mu, T)$ be a measure-preserving system. Then the (μ-a.e. determined) *Hopf decomposition* of X with respect to T refers to the decomposition into the *conservative part*

$$C_T := \left\{ x \in X : \sum_{k=0}^\infty g \circ T^k(x) = \infty \right\}$$

for some $g \in L_1(\mu)$ such that $g > 0$ and the *dissipative part* $D_T := X \setminus C_T$.

By Lemma 2.2.7 the uniqueness of $C_T \mod \mu$ is guaranteed for this decomposition and, as we will shortly see, we further have for every measure-preserving transformation T that $X = C_T \mod \mu$ is equivalent to T being conservative (cf. Corollary 2.2.11).

Proposition 2.2.9. *Let $(X, \mathcal{B}, \mu, T)$ be a measure-preserving system. Then:*
(a) *All wandering sets are subsets (mod μ) of the dissipative part D_T of X.*
(b) *For all measurable subsets A of the dissipative part D_T with $\mu(A) > 0$ there exists a wandering set $W \subset A$ with $\mu(W) > 0$.*

Proof. To prove (a) fix $f \in L_1(\mu)$ with $f > 0$ and let $W \in \mathcal{W}_T$ with $\mu(W) > 0$. Then, for every $n \in \mathbb{N}$, we have

$$\int_W \sum_{k=0}^n f \circ T^k \, d\mu = \sum_{k=0}^n \int \mathbb{1}_W \cdot f \circ T^{n-k} \, d\mu$$

$$= \sum_{k=0}^n \int \mathbb{1}_W \circ T^k \cdot f \circ T^n \, d\mu$$

$$= \int f \circ T^n \cdot \sum_{k=0}^n \mathbb{1}_W \circ T^k \, d\mu \leq \int f \, d\mu < \infty.$$

Since we suppose that the measure of W is positive, we have that the infinite sum $\sum_{k=0}^{\infty} f \circ T^k$ must converge μ almost surely on W. Hence, W is contained in the dissipative part.

Towards part (b) fix a measurable subset A of the dissipative part with positive measure and without loss of generality we also assume $\mu(A)$ to be finite. If we make the assumption that $\mu(A \cap W) = 0$ for all wandering sets $W \in \mathcal{B}$, then Halmos's Recurrence Theorem (Theorem 2.2.6) would imply that $\sum_{n \in \mathbb{N}} \mathbb{1}_A \circ T^n = \infty$ almost everywhere on A. But then Lemma 2.2.7 with $g := \mathbb{1}_A$, would imply that A is a subset of the conservative part. This contradiction finishes the proof. □

Remark 2.2.10. What we have just proved is that the dissipative part D_T of a measure-preserving system is the *measurable union* of $\mathcal{W}_T$. This means by definition that the collection $\mathcal{W}_T$ is hereditary (i.e., measurable subsets of wandering sets are wandering sets) and that the properties of $\mathcal{W}_T$ described in (a) (that $\mathcal{W}_T$ is said to *cover* D_T) and (b) (that $\mathcal{W}_T$ is said to *saturate* D_T) are fulfilled. In fact, any measurable set with these properties is uniquely determined mod μ. To see why, suppose there were two measurable unions D and D' such that $\mu(D \setminus D') > 0$. Then by property (b) there exists a wandering set $W \subset D \setminus D'$ with positive measure and by property (a) this set must lie in D' which gives a contradiction. The same argument with D and D' interchanged gives that $D = D'$ mod μ.

The measurable union for a hereditary family of measurable sets (like $\mathcal{W}_T$) always exists and can also be constructed abstractly.

Corollary 2.2.11. *Let $(X, \mathcal{B}, \mu, T)$ be a measure-preserving system. Then the system is conservative if and only if*

$$C_T = X \quad \text{mod } \mu.$$

Proof. If the system is conservative, then by definition all elements of $\mathcal{W}_T$ have measure zero. Hence, by Proposition 2.2.9 (b) we have $\mu(D_T) = 0$.

Conversely, if $C_T = X$ mod μ then $\mu(D_T) = 0$. Hence, by Proposition 2.2.9 (a) the system must be conservative. □

The next theorem provides a convenient way of determining whether a given transformation is conservative with respect to an infinite invariant measure (cf. Remark 2.3.22 (3) for an application). We will use the following notion of a sweep-out set.

Definition 2.2.12. Let $(X, \mathcal{B}, \mu)$ be a measure space and $T : X \to X$ a map. A set $A \in \mathcal{B}$ is called a *sweep-out set* for T if A is of finite, positive μ-measure and

$$\bigcup_{n=0}^{\infty} T^{-n}(A) = X \quad \text{mod } \mu.$$

Remark 2.2.13. Note that for any sweep-out set A we have

$$\bigcup_{k \geq n} T^{-k}A = T^{-n}\bigcup_{k \geq 0} T^{-k}A = T^{-n}X = X \quad \text{mod } \mu,$$

for all $n \in \mathbb{N}$, and hence

$$\bigcap_{n \in \mathbb{N}} \bigcup_{k \geq n} T^{-k}A = X \quad \text{mod } \mu.$$

Theorem 2.2.14 (Maharam's Recurrence Theorem). *Let $(X, \mathcal{B}, \mu, T)$ be a measure-preserving system and suppose that there exists a sweep-out set A for T. Then T is conservative.*

Proof. This theorem is a direct consequence of Lemma 2.2.7 and Corollary 2.2.11 since for $g := \mathbb{1}_A \in L_1^+(\mu)$ and $f \in L_1(\mu)$ with $f > 0$, arbitrary, we obtain with the help of Remark 2.2.13

$$X = \left\{ x \in X : \sum_{k=0}^{\infty} \mathbb{1}_A(T^k(x)) = \infty \right\} \subset \left\{ x \in X : \sum_{k=0}^{\infty} f(T^k(x)) = \infty \right\} \quad \text{mod } \mu. \qquad \square$$

Let us end this section with another recurrence theorem, this one giving information about distances of orbits mapped by a measurable function to a metric space.

Theorem 2.2.15 (Poincaré's Recurrence Theorem). *Let $T : X \to X$ be a conservative non-singular transformation of a σ-finite measure space $(X, \mathcal{B}, \mu)$, let (M, d) be a separable metric space and $f : X \to M$ a measurable function. Then μ-a.e. we have*

$$\liminf_{n \to \infty} d\left(f(x), f \circ T^n(x)\right) = 0.$$

Proof. Let $B \subset M$ be a measurable set of diameter less than $1/n$. Then by Halmos's Recurrence Theorem 2.2.6 we have μ-a.e. on $f^{-1}B$

$$\sum_{n=0}^{\infty} \mathbb{1}_{f^{-1}B} \circ T^n = \infty.$$

Consequently, there exists a μ-null set $N_B \in \mathcal{B}$ such that for all $x \in f^{-1}B \setminus N_B$ we have

$$\liminf_{n \to \infty} d\left(f(x), f \circ T^n(x)\right) \leq \frac{1}{n}.$$

By the separability of M there exists up to measure zero a countable cover of X by sets of the form $f^{-1}B \setminus N_B$ with measurable subsets $B \subset M$ of diameter less than $1/n$. The union of these sets U_n has full measure and for all $x \in U_n$ we have

$$\liminf_{n \to \infty} d\left(f(x), f \circ T^n(x)\right) \leq \frac{1}{n}.$$

The claim of the theorem then holds for all x in $\bigcap_{n \in \mathbb{N}} U_n$, which is a set of full measure. $\qquad\square$

2.3 The transfer operator

2.3.1 Jacobians and the change of variable formula

Now, we remind the reader of absolutely continuous measures and the Radon–Nikodým Theorem, which will be utilised heavily for the rest of this section.

Definition 2.3.1. Let μ and ν be two measures on a measurable space $(X, \mathcal{B})$. Then ν is called *absolutely continuous with respect to μ*, or sometimes μ-absolutely continuous, and written $\nu \ll \mu$, if for all $B \in \mathcal{B}$ with $\mu(B) = 0$ we have that $\nu(B) = 0$. Moreover, if for the two measures μ and ν we have that $\nu \ll \mu$ as well as $\mu \ll \nu$, then the two measures are said to be *equivalent* and we write $\mu \sim \nu$. We will also refer to equivalent measures as being in the same *measure class*.

Note that the definition of a non-singular transformation given in Definition 2.2.4 can also be phrased in the following way: The transformation $T : X \to X$ is non-singular if the measure $\mu \circ T^{-1}$ is absolutely continuous with respect to μ.

Remark 2.3.2. Let us point out here that Proposition 2.1.12 shows that the Gauss measure m_G is absolutely continuous with respect to the Lebesgue measure λ and vice versa. Hence, the two measures are equivalent.

Lemma 2.3.3. *Let $(X, \mathcal{B}, \mu)$ be a σ-finite measure space and $g \in \mathcal{M}^+(\mathcal{B})$. Then the integral*

$$\mu_g(A) := \int_A g \, d\mu, \ A \in \mathcal{B}, \tag{2.3}$$

defines a σ-finite measure μ_g on $(X, \mathcal{B})$, which is absolutely continuous with respect to μ. We have $g \in L_1^+(\mu)$ if and only if μ_g is finite. We have $\mu_g = \mu_h$ for two measurable non-negative functions g, h if and only if $g \sim h$, in other words, if $g = h$ as elements of $\mathcal{M}^+(\mathcal{B})$.

The unique function $g \in \mathcal{M}^+(\mathcal{B})$ with $\nu = \mu_g$ is called the density *of ν with respect to μ.*

Proof. The fact that μ_g is a σ-additive set function (and hence defines a measure) follows from the Monotone Convergence Theorem. Since μ is σ-finite there exists a sequence of measurable sets $B_1 \subset B_2 \subset \cdots$ such that $\bigcup B_k = X$ and $\mu(B_k) < \infty$. Since $g < \infty$ on X we have that $A_k := B_k \cap \{g \le k\}$ defines an increasing sequence of sets with union equal to X for which we have $\mu_g(A_k) \le k\mu(B_k) < \infty$. This shows that also μ_g is σ-finite. Clearly, $\mu_g \ll \mu$ and $g \in L_1^+(\mu)$ if and only if μ_g is finite.

If $f \sim g$ then it follows immediately that $\mu_f = \mu_g$. For the reverse implication define $C_k := B_k \cap \{f < g\}$. Then, since $\int_{C_k} g - f \, d\mu = \mu_g(C_k) - \mu_f(C_k) = 0$ and the integrand is non-negative, it follows that $f = g \, \mu$-a.e. on C_k. Analogously, one shows that $f = g \, \mu$-a.e. on $B_k \cap \{f \geq g\}$. Hence, $f = g \, \mu$-a.e. on B_k for every $k \in \mathbb{N}$ and consequently $f \sim g$. □

The significance of the Radon–Nikodým Theorem is that also the converse of Lemma 2.3.3 holds, that is, every finite measure ν such that $\nu \ll \mu$ has a density with respect to μ.

Theorem 2.3.4 (The Radon–Nikodým Theorem). *Let μ and ν be two σ-finite measures on a measurable space $(X, \mathcal{B})$ such that $\nu \ll \mu$. Then there exists a unique element $h \in \mathcal{M}^+(\mathcal{B})$ such that $\nu = \mu_h$, that is for every set $B \in \mathcal{B}$, we have*

$$\nu(B) = \int_B h \, d\mu.$$

Moreover, if the measure ν is finite, then this almost surely unique function h belongs to $L_1(\mu)$.

Proof. See, for instance, Theorem 6.10 in Rudin [Rud87], (where the theorem is proved in slightly greater generality than stated here). □

Remark 2.3.5. The unique density h appearing in Theorem 2.3.4 is often referred to as the *Radon–Nikodým derivative* of ν with respect to μ and denoted by $h = d\nu/d\mu$.

2.3.2 Obtaining invariant measures via the transfer operator

Our aim now is to obtain invariant measures which are absolutely continuous to a given reference measure. Throughout, we consider a non-singular transformation $T : X \to X$ on a σ-finite measure space $(X, \mathcal{B}, \mu)$. Let $g \in \mathcal{M}^+(\mathcal{B})$ and suppose that μ_g has density g with respect to μ, that is for all measurable sets A we have $\mu_g(A) := \int_A g \, d\mu$. Then, since $\mu_g \ll \mu$, we have $\mu_g \circ T^{-1} \ll \mu \circ T^{-1} \ll \mu$. Thus, via the Radon–Nikodým Theorem, we can define the operator $\widehat{T}_\mu : \mathcal{M}^+(\mathcal{B}) \to \mathcal{M}^+(\mathcal{B})$ by

$$\widehat{T}_\mu(g) := \frac{d(\mu_g \circ T^{-1})}{d\mu}.$$

If it is clear which measure μ is meant, we will simply write $\widehat{T} := \widehat{T}_\mu$. If $g \in L_1(\mu)$, then we extend this definition linearly by setting $\widehat{T}(g) := \widehat{T}(g^+) - \widehat{T}(g^-)$. Defined in this way, $\widehat{T} : L_1(\mu) \to L_1(\mu)$ is a positive, bounded, linear operator. In fact, for the operator norm of $\widehat{T}$, we have that $\|\widehat{T}\| = 1$ (the proof of this fact is left to Exercise 2.6.11).

Definition 2.3.6. The operator $\widehat{T}_\mu : L_1(\mu) \to L_1(\mu)$ defined above is called the *transfer operator* of T with respect to the measure μ.

Let us make one further observation. For all $A \in \mathcal{B}$, we have that

$$\int_X \mathbb{1}_A \cdot \widehat{T}(g) \, d\mu = \int_A \frac{d(\mu_g \circ T^{-1})}{d\mu} \, d\mu = \mu_g \circ T^{-1}(A)$$

$$= \int_{T^{-1}(A)} g \, d\mu = \int_X (\mathbb{1}_A \circ T) \cdot g \, d\mu.$$

Then, an approximation argument shows that for all $f \in L_\infty(\mu)$, we equivalently have that

$$\int_X f \cdot \widehat{T}(g) \, d\mu = \int_X (f \circ T) \cdot g \, d\mu. \tag{2.4}$$

Inductively, it follows for all $k \in \mathbb{N}$ that $\int_X f \cdot \widehat{T}^k(g) \, d\mu = \int_X (f \circ T^k) \cdot g \, d\mu$. Furthermore, the relation in (2.4) characterises $\widehat{T}(g)$. Indeed, suppose there exist g_1 and g_2 in $L_1(\mu)$ such that for all $f \in L_\infty(\mu)$ we have

$$\int f g_1 \, d\mu = \int (f \circ T) g \, d\mu = \int f g_2 \, d\mu.$$

Then if we choose $f := \mathrm{sgn}(g_1 - g_2)$, it follows that

$$\int |g_1 - g_2| \, d\mu = \int f(g_1 - g_2) \, d\mu = \int f g_1 \, d\mu - \int f g_2 \, d\mu = 0,$$

and thus $g_1 \sim g_2$.

Remark 2.3.7. The relation in (2.4) shows that the transfer operator captures the evolution of probability densities under the action of $T : [0, 1] \to [0, 1]$ in the following sense. Suppose that X_0 denotes a $[0, 1]$-valued random variable with a distribution absolutely continuous with respect to μ and density g. That is, $\mathbb{P}(X_0 \in A) = \int_A g \, d\mu$, for all $A \in \mathcal{B}$. Then, for all $n \in \mathbb{N}$, the distribution of the random variable $X_n := T^n \circ X_0$ also has a density with respect to μ, and that density coincides with $\widehat{T}^n g$.

Remark 2.3.8. For a σ-finite measure μ, it is a fact that $(L_1(\mu))^* \simeq L_\infty(\mu)$. (Here the star denotes the dual space of $L_1(\mu)$, where we recall that if X is a normed linear space, then the dual space X^* is defined to be the set of all continuous linear functionals $f : X \to \mathbb{R}$.) Hence, the operator

$$U_T : L_\infty(\mu) \to L_\infty(\mu),$$
$$f \mapsto f \circ T$$

is the adjoint operator of $\widehat{T}$ (Exercise 2.6.10). The operator U_T is usually referred to as the *Koopman operator* (see [Koo31]).

We now formulate a dual version of Proposition 2.2.9. For this fix some $f \in L_1(\mu)$ with $f > 0$ and define the set $\widehat{C}_T$ and its complement $\widehat{D}_T$ by setting

$$\widehat{C}_T := \left\{ \sum_{k \in \mathbb{N}} \widehat{T}^k f = \infty \right\} \quad \text{and} \quad \widehat{D}_T := \left\{ \sum_{k \in \mathbb{N}} \widehat{T}^k f < \infty \right\}.$$

This decomposition gives rise to a generalisation of the previously-defined Hopf decomposition for measure-preserving systems (see Remark 2.3.10).

Proposition 2.3.9. *Let $(X, \mathcal{B}, \mu, T)$ be a non-singular system. Then:*
(a) *All wandering sets are subsets of $\widehat{D}_T$ mod μ.*
(b) *For all measurable subsets $A \subset \widehat{D}_T$ of the dissipative part with $\mu(A) > 0$ there exists a wandering set $W \in \mathcal{B}$ with $\mu(W) > 0$ and $W \subset A$.*

Proof. Towards part (a), let $W \in \mathcal{B}$ be a wandering set, that is $\sum_{n=0}^{\infty} \mathbb{1}_W \circ T^n \le 1$, μ-a.e. Then we have

$$\int_W \sum_{k \in \mathbb{N}} \widehat{T}^k f \, d\mu = \int f \cdot \sum_{k \in \mathbb{N}} \mathbb{1}_W \circ T^k \, d\mu \le \int f \, d\mu < \infty.$$

That the integral is finite implies that the sum $\sum_{k \in \mathbb{N}} \widehat{T}^k f$ is finite μ almost everywhere on W. This finishes the proof of part (a).

To prove (b), first fix a measurable subset

$$A \subset \bigcup_{N \in \mathbb{N}} \left\{ \sum_{k \in \mathbb{N}} \widehat{T}^k f \le N \right\} = D_{\widehat{T}}$$

with positive measure. Then there exists a measurable subset $B \subset A \cap \left\{ \sum_{k \in \mathbb{N}} \widehat{T}^k f \le N \right\}$ for some $N \in \mathbb{N}$ with positive and finite measure.

Now, if we assume that $\mu(A \cap W) = 0$ for all wandering sets $W \in \mathcal{B}$ then Halmos's Recurrence Theorem 2.2.6 would imply that $\sum_{n \in \mathbb{N}} \mathbb{1}_B \circ T^n = \infty$ almost everywhere on B. But then we would have

$$\infty > N\mu(B) \ge \int \sum_{k \in \mathbb{N}} \widehat{T}^k f \cdot \mathbb{1}_B \, d\mu = \int f \sum_{k \in \mathbb{N}} \mathbb{1}_B \circ T^k \, d\mu = \infty.$$

This contradiction finishes the proof. □

Remark 2.3.10. The last proposition shows in particular that for a non-singular dynamical system, the set $\widehat{D}_T$ is the measurable union of the wandering sets (cf. Remark 2.2.10). Hence, this decomposition is
- independent of the chosen positive integrable function f and
- for a measure-preserving system coincides with the previously-defined Hopf decomposition.

The following corollary allows us to characterise conservativity also in terms of the transfer operator.

Corollary 2.3.11. *Let $(X, \mathcal{B}, \mu, T)$ be a non-singular dynamical system. Then the system is conservative if and only if*

$$\widehat{C}_T = X \mod \mu.$$

Proof. The proof follows exactly along the lines of the proof of Corollary 2.2.11. □

Example 2.3.12. Let us consider the extended tent map $T : \mathbb{R} \to \mathbb{R}$ given by

$$T(x) := \begin{cases} 2x & \text{for } x \leq 1/2 \\ -2x + 2 & \text{for } x > 1/2. \end{cases}$$

which defines a non-singular transformation with respect to the Lebesgue measure λ. Then the conservative part of the Hopf decomposition of $\mathbb{R}$ is given by $\widehat{C}_T = [0, 1]$.

As advertised at the beginning of this section, we are now in a position to find a T-invariant measure with the help of the transfer operator.

Theorem 2.3.13. *Let $(X, \mathcal{B}, \mu, T)$ be a non-singular system. If $g \in \mathcal{M}^+(\mathcal{B})$ with $\widehat{T}_\mu(g) = g$, then the measure μ_g (as defined in (2.3)) is T-invariant and $\widehat{T}_{\mu_g}(\mathbb{1}) = \mathbb{1}$.*

Proof. Using (2.4), for all $B \in \mathcal{B}$ we have

$$\mu_g \circ T^{-1}(B) = \int \mathbb{1}_B \circ T \cdot g \, d\mu = \int \mathbb{1}_B \cdot \widehat{T}g \, d\mu = \int \mathbb{1}_B g \, d\mu = \mu_g(B).$$

From this we deduce that

$$\widehat{T}_{\mu_g}(\mathbb{1}) = \frac{d(\mu_g \circ T^{-1})}{d\mu_g} = \frac{d\mu_g}{d\mu_g} = \mathbb{1}.$$

□

The reason for naming $\widehat{T}$ the transfer operator should now be clear. The idea behind this operator is that first of all it transfers the action of T on X to an action on $L_1(\mu)$ and secondly, it transfers in this way the measure-theoretic problem of finding a T-invariant measure to the functional-analytic problem of finding a fixed point for the operator $\widehat{T}$. This is all very well, but without an explicit formula for $\widehat{T}$, we are still not really any closer to actually writing down an invariant measure for our remaining examples. In the following section, we will address this issue.

2.3.3 The Ruelle operator

Let us now specialize somewhat to the case that T is a continuous map of the circle $\mathbb{R}/\mathbb{Z}$ such that T admits a Markov partition $\{A_i : i \geq 1\}$, as in Definition 1.2.21. Let us further assume that the map T consists of full branches, that is, that $T|_{A_i}(A_i) = (0, 1)$ for all $i \geq 1$. We are interested in finding invariant measures that are absolutely continuous with respect to the Lebesgue measure λ, since these retain some physical meaning, so instead of the transfer operator associated to some arbitrary σ-finite measure μ, we shall consider the special case of the transfer operator with respect to λ.

Remark 2.3.14. If $\mu \sim \lambda$ and $h := d\mu/d\lambda$ denotes the a.e. positive density of μ with respect to λ (see Exercise 2.6.13), then the operators $\widehat{T}_\mu$ and $\widehat{T}_\lambda$ are related as follows:

$$\widehat{T}_\mu(f) = \frac{1}{h}\widehat{T}_\lambda(h \cdot f).$$

Indeed, for every $A \in \mathcal{B}$, using the identity in (2.4) twice, first for the measure μ and then for λ, we obtain that

$$\int_A \widehat{T}_\mu(f)\, d\mu = \int_{[0,1]} \mathbb{1}_A \cdot \widehat{T}_\mu(f)\, d\mu = \int_{[0,1]} (\mathbb{1}_A \circ T) \cdot f\, d\mu$$

$$= \int_{[0,1]} (\mathbb{1}_A \circ T) \cdot fh\, d\lambda = \int_{[0,1]} \mathbb{1}_A \cdot \widehat{T}_\lambda(h \cdot f)\, d\lambda$$

$$= \int_A \frac{1}{h}\widehat{T}_\lambda(h \cdot f)\, d\mu.$$

If $g \in \mathcal{M}^+(\mathcal{B})$ is such that $\widehat{T}_\lambda(g) = g$, then $d\lambda_g := g\, d\lambda$ defines a T-invariant measure absolutely continuous with respect to λ and hence $\widehat{T}_{\lambda_g}\mathbb{1} = \mathbb{1}$.

Our aim here, as advertised at the end of the last section, is to define another, related, operator which has a more concrete form. For this, we will need the following lemma.

Lemma 2.3.15. Let $\varphi : (a, b) \to (0, 1)$ be continuously differentiable such that either $\varphi' > 0$ or $\varphi' < 0$. Then we have

$$\frac{d\lambda \circ \varphi^{-1}}{d\lambda} = |(\varphi^{-1})'|.$$

Proof. We only consider the case $\varphi' > 0$. Let $g := \mathbb{1}_J$ for some subinterval J of $(0, 1)$. Then by the substitution rule of integration we have

$$\lambda \circ \varphi^{-1}(J) = \int g \circ \varphi\, d\lambda = \int g \circ \varphi \cdot \varphi'/\varphi'\, d\lambda = \int g \circ \varphi \cdot (\varphi^{-1})' \circ \varphi \cdot \varphi'\, d\lambda$$

$$= \int g \cdot (\varphi^{-1})'\, d\lambda = \int_J (\varphi^{-1})'\, d\lambda.$$

Since the subintervals of $(0, 1)$ generate $\mathcal{B}$ we have verified the claim of the lemma. $\square$

For an interval map T, as described above, we have that the maps $T|_{A_i}$ are invertible with measurable inverse branches $T_i := (T|_{A_i})^{-1}$. Further, suppose that $\lambda \circ T_i \ll \lambda$.

Hence, we can define the Jacobian for each inverse branch $i \geq 1$ to be

$$\mathfrak{J}_i := \frac{d(\lambda \circ T_i)}{d\lambda}.$$

Proposition 2.3.16. *For each $g \in \mathcal{M}^+(\mathcal{B})$ or $g \in L_1(\mu)$, we have*

$$\widehat{T}_\lambda(g) = \sum_{i \geq 1} g \circ T_i \cdot \mathfrak{J}_i.$$

Proof. Let g be as assumed and let $A \in \mathcal{B}$ be arbitrary. Then,

$$\lambda_g \circ T^{-1}(A) = \lambda_g \left(\bigcup_{i \geq 1} T_i(A) \right) = \sum_{i \geq 1} \lambda_g((T_i(A)))$$

$$= \sum_{i \geq 1} \int \mathbb{1}_A \circ T_i^{-1} \cdot g \circ T_i \circ T_i^{-1} \, d\lambda = \sum_{i \geq 1} \int \mathbb{1}_A \cdot g \circ T_i \, d(\lambda \circ T_i)$$

$$= \int \mathbb{1}_A \sum_{i \geq 1} g \circ T_i \cdot \mathfrak{J}_i \, d\lambda.$$

Since the density is unique this proves the proposition. □

Definition 2.3.17. Suppose that the dynamical system $([0, 1], T)$ is such that all the inverse branches $T_i := (T|_{A_i})^{-1}$ are additionally continuously differentiable. Then the *Ruelle operator* P_T is defined for any function $f : [0, 1] \to \mathbb{R}$ such that the right-hand side makes sense for $x \in (0, 1)$ to be

$$P_T(f)(x) := \sum_{i \geq 1} (f \circ T_i)(x) \cdot |T_i'(x)|. \tag{2.5}$$

Proposition 2.3.18. *We have that P_T is well-defined as an operator on $L_1(\lambda)$. When we use the same notation for P_T acting on functions and acting on equivalence classes, we also have that*

$$P_T = \widehat{T}_\lambda.$$

Proof. We have that if f, g are two integrable functions in the same $L_1(\lambda)$ equivalence class, then also

$$P_T(f) = P_T(g), \quad \lambda\text{-almost everywhere.}$$

To see this, it suffices to note that under strictly monotone continuously differentiable functions, Lebesgue null-sets are mapped to null-sets.

Since by Lemma 2.3.15 we have $\mathfrak{J}_i = |T_i'|$ the second statement follows from Proposition 2.3.16. □

2.3.4 Invariant measures for F, F_α, G and L_α

The above discussion allows us to finally get our hands on Lebesgue-absolutely continuous invariant measures for F and F_α; we can also revisit the invariant measures for G and L_α in terms of the relevant transfer operators. In light of Proposition 2.3.18, we have that if we find a fixed point of the Ruelle operator, this gives us a fixed point of the transfer operator.

Proposition 2.3.19. *For the Farey system* $([0, 1], \mathcal{B}, F)$*, the λ-absolutely continuous measure ν_F defined by the density h_F, given by $h_F(x) := 1/x$, is an F-invariant measure. Moreover, the measure ν_F is infinite and σ-finite.*

Proof. First observe that according to (2.5), the Ruelle operator P_F for the map F acts in the following way: For $f : [0, 1] \to \mathbb{R}$ measurable,

$$P_F(f) = |F_0'| \cdot (f \circ F_0) + |F_1'| \cdot (f \circ F_1),$$

where $F_0(x) := x/(1 + x)$ and $F_1(x) := 1/(1 + x)$ denote the inverse branches of the Farey map (as defined in Section 1.3.1). In order to show that the density h_F defines an F-invariant measure absolutely continuous with respect to λ, it suffices to show that $P_F(h_F) = h_F$. Indeed, we have

$$F_0'(x) = \frac{1}{(1 + x)^2} \quad \text{and} \quad F_1'(x) = \frac{-1}{(1 + x)^2}.$$

The following calculation then finishes the proof:

$$P_F(h_F)(x) = \frac{1}{(1 + x)^2} \left(h_F\left(\frac{x}{1 + x}\right) + h_F\left(\frac{1}{1 + x}\right) \right)$$
$$= \frac{(1 + x) + x(1 + x)}{x(1 + x)^2} = \frac{1}{x} = h_F(x). \qquad \square$$

Before moving on to consider invariant measures for the α-Farey systems, let us make an interesting observation about fractions in the Stern–Brocot tree (which was introduced directly after Definition 1.3.4). For each reduced fraction $v/w \in (0, 1)$ and each $n \in \mathbb{N}_0$, we have that

$$\sum_{p/q \in F^{-n}(v/w)} \frac{1}{pq} = \frac{1}{vw}. \tag{2.6}$$

In particular, for the special case $v/w = 1/2$, we have that

$$\sum_{p/q \in F^{-n}(1/2)} \frac{2}{pq} = 1.$$

The identity in this special case was first noted by the Canadian music theorist Pierre Lamothe (see reference in [Gut11]), whereas the general case originates in [KS12a]. To first prove (2.6) in an elementary way, suppose that the statement is true for all reduced fractions $v/w \in (0, 1)$ and for some $n \in \mathbb{N}_0$. Then,

$$\sum_{\frac{p}{q} \in F^{-(n+1)}(\frac{v}{w})} \frac{1}{pq} = \sum_{\frac{p}{q} \in F^{-n}(F^{-1}(\frac{v}{w}))} \frac{1}{pq}$$

$$= \sum_{\frac{p}{q} \in F^{-n}(\frac{v}{v+w})} \frac{1}{pq} + \sum_{\frac{p}{q} \in F^{-n}(\frac{w}{v+w})} \frac{1}{pq}$$

$$= \frac{1}{v(v+w)} + \frac{1}{w(v+w)} = \frac{1}{vw}.$$

Alternatively, the equality in (2.6) can be deduced as a special case from the fixed point equation for the Ruelle operator P_F for the map F. As in Proposition 2.3.19, let $h_F(x) := 1/x$ denote the density which satisfies $P_F(h_F) = h_F$. For all $x \in (0, 1)$ and all $n \in \mathbb{N}_0$, we then have that

$$\sum_{y \in F^{-n}(x)} |(F^n)'(y)|^{-1} h_F(y) = h_F(x). \tag{2.7}$$

(You are asked to prove the above statement in Exercise 2.6.12.) A straightforward calculation, which we leave to Exercise 2.6.14, shows that

$$\left|(F^n)'\left(\frac{p}{q}\right)\right| = \frac{q^2}{w^2}, \quad \text{for all } \frac{p}{q} \in F^{-n}\left(\frac{v}{w}\right).$$

Inserting this into (2.7), it follows that

$$\sum_{\frac{p}{q} \in F^{-n}(\frac{v}{w})} \frac{w^2}{pq} = \sum_{\frac{p}{q} \in F^{-n}(\frac{v}{w})} \frac{w^2}{q^2} \cdot \frac{q}{p}$$

$$= \sum_{\frac{p}{q} \in F^{-n}(\frac{v}{w})} \left|(F^n)'\left(\frac{p}{q}\right)\right|^{-1} h_F\left(\frac{p}{q}\right) = h_F\left(\frac{v}{w}\right) = \frac{w}{v},$$

and hence the statement in (2.6) follows.

Proposition 2.3.20. *For each α-Farey system $([0, 1], \mathcal{B}, F_\alpha)$, the λ-absolutely continuous measure ν_α, given by the density h_{F_α}, which is defined, up to multiplication by a constant, by*

$$h_{F_\alpha} := \frac{d\nu_\alpha}{d\lambda} = \sum_{n=1}^{\infty} \frac{t_n}{a_n} \cdot \mathbb{1}_{A_n},$$

is an F_α-invariant measure. Moreover, ν_α is σ-finite, and we have that ν_α is an infinite measure if and only if α is of infinite type.

Proof. We will prove this as in the case of the Farey map by considering the Ruelle operator P_{F_α} for the map F_α, which acts on measurable functions $f : [0, 1] \to \mathbb{R}$ by

$$P_{F_\alpha}(f) = |F_{\alpha,0}'| \cdot (f \circ F_{\alpha,0}) + |F_{\alpha,1}'| \cdot (f \circ F_{\alpha,1}).$$

As in Proposition 2.3.19, it is sufficient to show that h_{F_α} is an eigenfunction of P_{F_α}, that is,

$$P_{F_\alpha}(h_{F_\alpha}) = h_{F_\alpha}.$$

To see this, note that for the inverse branches $F_{\alpha,1}$ and $F_{\alpha,0}$ and the density h_{F_α}, an easy computation shows that we have

$$h_{F_\alpha} \circ F_{\alpha,1} = t_1/a_1 \cdot \mathbb{1}_{[0,1]} \text{ and } h_{F_\alpha} \circ F_{\alpha,0} = \sum_{n=1}^{\infty} t_{n+1}/a_{n+1} \cdot \mathbb{1}_{A_n}.$$

Moreover, one immediately verifies that

$$|F_{\alpha,1}'| = a_1 \cdot \mathbb{1}_{[0,1]} \text{ and } |F_{\alpha,0}'| = \sum_{n=1}^{\infty} a_{n+1}/a_n \cdot \mathbb{1}_{A_n}.$$

These two observations together imply that

$$P_{F_\alpha}(h_{F_\alpha}) = |F_{\alpha,0}'| \cdot (h_{F_\alpha} \circ F_{\alpha,0}) + |F_{\alpha,1}'| \cdot (h_{F_\alpha} \circ F_{\alpha,1})$$

$$= \sum_{n=1}^{\infty} \left(\frac{a_{n+1}}{a_n} \frac{t_{n+1}}{a_{n+1}} \right) \cdot \mathbb{1}_{A_n} + t_1 \cdot \mathbb{1}_{[0,1]}$$

$$= \sum_{n=1}^{\infty} \left(\frac{t_{n+1}}{a_n} + 1 \right) \cdot \mathbb{1}_{A_n} = \sum_{n=1}^{\infty} \frac{t_n}{a_n} \cdot \mathbb{1}_{A_n} = h_{F_\alpha}.$$

Regarding the second statement of the proposition, the σ-finiteness of the measure ν_α follows from the fact that $\nu_\alpha(A_n) < \infty$ for all $n \in \mathbb{N}$, and a simple calculation shows that for each $n \in \mathbb{N}$ we have that

$$\nu_\alpha \left(\bigcup_{k=1}^{n} A_k \right) = \sum_{k=1}^{n} \nu_\alpha(A_k) = \sum_{k=1}^{n} \int_{A_k} h_{F_\alpha} \, d\lambda = \sum_{k=1}^{n} \frac{t_k}{a_k} \cdot a_k = \sum_{k=1}^{n} t_k.$$

Recalling that α is of infinite type provided that $\sum_{k=1}^{\infty} t_k$ diverges, the proof is finished. □

To complete the picture for our main examples, we recall the following facts for the Gauss and the α-Lüroth system.

Proposition 2.3.21.

(a) *For the Gauss system $([0, 1], \mathcal{B}, G)$, the λ-absolutely continuous measure m_G, given by the density h_G, which is defined by*

$$h_G(x) := \frac{dm_G}{d\lambda} = \frac{1}{\log 2} \frac{1}{1 + x},$$

is a G-invariant probability measure.

(b) *For an α-Lüroth system $([0, 1], \mathcal{B}, L_\alpha)$, the λ-absolutely continuous measure m_α, given by the density h_{L_α}, which is defined by*

$$h_{L_\alpha}(x) := \frac{dm_\alpha}{d\lambda} = \mathbb{1}_{[0,1]},$$

is a L_α-invariant probability measure. In other words, m_α coincides with λ.

Proof. The statements in (a) and (b) have already been obtained in Proposition 2.1.10 and Proposition 2.1.8, respectively. However, for the Gauss map G this statement can now be derived alternatively by considering the Ruelle operator P_G for G, which is defined, for measurable functions $f : [0, 1] \to \mathbb{R}$ and $x \in [0, 1]$, by

$$P_G(f)(x) := \sum_{n=1}^{\infty} |G_n'(x)| f(G_n(x)),$$

with G_n referring to the n-th inverse branch of G (see also Remark 2.1.11). One then immediately verifies that $P_G(h_G) = h_G$, from which the assertion follows. The proof for L_α, using the Ruelle operator P_{L_α} for L_α, is left as an exercise (see Exercise 2.6.4). □

Remark 2.3.22.

1. It is clear that the measures ν_F, ν_α, m_G and m_α are all absolutely continuous with respect to λ; indeed, they are defined that way. The converse is also true, since their densities are strictly positive. Hence, all these measures are equivalent to λ.

2. The reader may have encountered the Bogolyubov–Krylov Theorem, which states that for an arbitrary continuous map $T : X \to X$ on a compact metric space X there always exists a T-invariant probability measure. However, this theorem does not say anything about whether the measure is absolutely continuous with respect to Lebesgue measure. There may be several invariant measures, but we will shortly see in Section 2.4.5 that in our leading examples the λ-absolutely continuous ones given by the densities h_F and h_{F_α} are unique for their respective systems.

3. We can use Maharam's Recurrence Theorem (see Theorem 2.2.14) to show that the Farey system $([0, 1], F, \nu_F)$ and any α-Farey system $([0, 1], F_\alpha, \nu_\alpha)$ is conservative. Indeed, this can be seen immediately on observing that both $\bigcup_{n=0}^{\infty} F^{-n}((1/2, 1])$ and $\bigcup_{n=0}^{\infty} F_\alpha^{-n}(A_1)$ are equal to $(0, 1]$.

2.3.5 Invariant measures via the jump transformation

Before moving on from the subject of invariant measures, let us discuss another way of obtaining an invariant measure for an infinite measure system, this time by way of the jump transformation. Recall that in Definitions 1.3.2 and 1.4.9, we introduced specific jump transformations for the maps F and F_α, respectively. We also proved that the Gauss map G is the jump transformation of the Farey map F with respect to the set $[1/2, 1]$ and the α-Lüroth map L_α is the jump transformation of the α-Farey map F_α with respect to the set A_1. Let us now give a more general definition.

Definition 2.3.23. Let $(X, \mathcal{B}, \mu, T)$ be a conservative measure-preserving system. Let E be a measurable set such that $T(E) = X$ and $\bigcup_{k \geq 0} T^{-k}E = X$. Then, the *first passage time* $p : X \to \mathbb{N}$ is defined to be

$$p(x) := 1 + \inf\{n \geq 0 : T^n(x) \in E\}$$

and the *jump transformation* $T_E^* : X \to X$ of T with respect to E is defined by setting

$$T_E^*(x) := T^{p(x)}(x).$$

To shorten the notation, let us write $\{p = n\}$ for the set $\{x \in X : p(x) = n\}$, so that $T_E^* = T^n$ on $\{p = n\}$. We also write $\{p > n\}$ for the set $\{x \in X : p(x) > n\}$. Observe that $\{p > 0\} = X$ and $\{p > n\} = \bigcup_{k=n+1}^{\infty} \{p = k\}$ for $n \geq 1$.

Example 2.3.24. In the case of the Farey map, setting $E := [1/2, 1]$ (so the jump transformation F_E^* coincides with the Gauss map, as was already shown earlier), yields that the sets $\{p = n\}$ are given by the first level Gauss cylinder sets, that is, for each $n \in \mathbb{N}$, we have that $\{p = n\} = C(n)$. Also, in this case we have that $\{p > 0\} = [0, 1]$ and $\{p > n\}$ is equal to the Farey cylinder with code consisting of n zeros, for $n \geq 1$.

One of the basic ideas behind the concept of the jump transformation is to find a set E for a given map T such that the jump transformation T_E^* is easier to understand than the original map T. The hope is to find a jump transformation which turns out to be a map that has already been studied earlier. The following lemma then yields information about T. (Throughout this section, it should be helpful to keep the Gauss map and the Farey map in mind, along with their invariant measures.)

Lemma 2.3.25. *With the notation above, assume that the map $T_E^* : X \to X$ preserves a finite measure ν. We then have that the measure μ, defined for any measurable set $B \in \mathcal{B}$ by*

$$\mu(B) := \sum_{n \geq 0} \nu(T^{-n}(B) \cap \{p > n\}),$$

is T-invariant.

Proof. Noting that $\{p > n\} = \{p > n+1\} \cup \{p = n+1\}$, we have that

$$\mu(T^{-1}(B)) = \sum_{n \geq 0} \nu(T^{-n}(T^{-1}(B)) \cap \{p > n\})$$

$$= \sum_{n \geq 0} \nu(T^{-n}(T^{-1}(B)) \cap \{p > n+1\})$$

$$+ \sum_{n \geq 0} \nu(T^{-n}(T^{-1}(B)) \cap \{p = n+1\})$$

$$= \sum_{n \geq 1} \nu(T^{-n}(B) \cap \{p > n\}) + \sum_{n \geq 1} \nu((T_E^*)^{-1}(B) \cap \{p = n\})$$

$$= \sum_{n \geq 0} \nu(T^{-n}(B) \cap \{p > n\}) = \mu(B).$$

$\square$

As an example, observe that if in the above lemma we put $B = E$, then we obtain that $\mu(E) = \nu(X)$. Also, let us consider the situation for the Farey and Gauss systems. We already know that the Gauss map G preserves the probability measure m_G. So, if we set $E := [1/2, 1]$, then Lemma 2.3.25 tells us that the Farey map F preserves the measure μ, where μ is given for arbitrary measurable $B \in \mathcal{B}$ by

$$\mu(B) := m_G(B) + \sum_{n \geq 1} m_G(F^{-n}(B) \cap \widehat{C}(0, \dots, 0)_n).$$

Here, $\widehat{C}(0, \dots, 0)_n$ denotes the level n Farey cylinder set with code consisting of n zeros, for each $n \geq 1$. To get an idea of what this measure actually looks like, let us calculate $\mu([1/(k+1), 1/k])$, for some $k \geq 2$ (note that for $k = 1$, this is precisely $\mu(E)$ and we already know that $\mu(E)$ is equal to $m_G(X)$). Observe first that $F^{-n}(B) \cap \widehat{C}(0, \dots, 0)_n$ is equal to $F_0^n(B)$, for any measurable set B. Therefore, we have that

$$\mu(B) = \sum_{n \geq 0} m_G(F_0^n(B)).$$

Recalling that $F_0(x) = x/(1+x)$, one immediately verifies that $F_0^n(x) = x/(1+nx)$. Thus,

$$\mu([1/(k+1), 1/k]) = \sum_{n \geq 0} m_G(F_0([1/(k+1), 1/k]))$$

$$= \sum_{n \geq 0} \frac{1}{\log 2} \int_{1/(k+n+1)}^{1/(k+n)} \frac{1}{1+x} \, d\lambda(x)$$

$$= \sum_{n \geq 0} \frac{1}{\log 2} \left(\log\left(1 + \frac{1}{k+n}\right) - \log\left(1 + \frac{1}{k+n+1}\right) \right)$$

$$= \frac{1}{\log 2} \log\left(\prod_{n \geq 0} \frac{k+n+1}{k+n} \cdot \frac{k+n+1}{k+n+2} \right)$$

$$= \frac{\log(k+1/k)}{\log 2}.$$

Recalling the measure ν_F obtained via the Ruelle operator in Proposition 2.3.19, we deduce that for the sets $[1/(k+1), 1/k]$, for $k \geq 1$,

$$\frac{1}{\log 2} \nu_F([1/(k+1), 1/k]) = \mu([1/(k+1), 1/k]).$$

It will turn out later that these two measures really are equal up to multiplication by a constant (see Theorem 2.4.35).

2.4 Ergodicity and exactness

Ergodicity is the natural way to describe indecomposability of transformations, in the sense that the system is said to be ergodic if and only if it is not possible to decompose it into two invariant subsystems supported on sets of positive measure.

Definition 2.4.1. The dynamical system $(X, \mathcal{B}, \mu, T)$ is said to be *ergodic* provided that whenever $A \in \mathcal{B}$ is such that $T^{-1}(A) = A$ we have that either $\mu(A) = 0$ or $\mu(X \setminus A) = 0$. In other words, an ergodic transformation has only trivial invariant subsets. We will often simply say that the map T is ergodic with respect to μ or that μ is an ergodic measure for T. If the system $(X, \mathcal{B}, \mu, T)$ is measure-preserving and T is ergodic, we will call the system an *ergodic measure-preserving system*.

Remark 2.4.2. Note that it is often useful to talk about T-invariant ergodic measures, and in many books ergodicity is only defined for T-invariant measures. However, this is not necessary. It can also be of interest to talk about ergodic measures that are not T-invariant.

Let us begin by considering the existence of sweep-out sets (see Definition 2.2.12) for conservative, ergodic systems.

Lemma 2.4.3. *If $(X, \mathcal{B}, \mu, T)$ is a conservative and ergodic non-singular system, then every measurable set of positive and finite measure is a sweep-out set for T.*

Proof. Suppose that $E \in \mathcal{B}$ is such that $\mu(E) > 0$. Further, let $E' := E \setminus W$, where $W := \{x \in E : T^n(x) \notin E \text{ for all } n \geq 1\}$. Then, by conservativity (see Halmos's Recurrence Theorem 2.2.6), we have that $\mu(W) = 0$ and thus, $\mu(E') > 0$. Now, noting that $x \in \bigcup_{n=0}^{\infty} T^{-n}(E')$ if and only if $x \in \bigcup_{n=1}^{\infty} T^{-n}(E')$, we have that

$$T^{-1}\left(\bigcup_{n=0}^{\infty} T^{-n}(E')\right) = \bigcup_{n=0}^{\infty} T^{-n}(E'),$$

and thus, since $\mu(\bigcup_{n=0}^{\infty} T^{-n}(E')) > 0$, the ergodicity of T implies that

$$\bigcup_{n=0}^{\infty} T^{-n}(E) = \bigcup_{n=0}^{\infty} T^{-n}(E') = X \quad \text{mod } \mu.$$

(Recall that the we use the notation $A = B \bmod \mu$ to mean that the two sets A and B are equal up to a set of μ-measure zero). $\qquad\square$

Corollary 2.4.4. *If $(X, \mathcal{B}, \mu, T)$ is a non-singular system, then T is conservative and ergodic if and only if for every measurable function $f : X \to [0, \infty)$ with $\int f \, d\mu > 0$ we have that*

$$\sum_{k=0}^{\infty} f \circ T^k = \infty, \quad \mu\text{-a.e. on } X.$$

Proof. Suppose first that the condition stated for measurable functions holds. Then let $A \in \mathcal{B}$ be either a wandering set or a T-invariant proper subset of X and suppose that $\mu(A) > 0$. It follows that

$$\left\{ \sum_{k=0}^{\infty} \mathbb{1}_A \circ T^k = \infty \right\} \neq X \quad \bmod \mu$$

contradicting our assumption for $f = \mathbb{1}_A$. Hence, such a set A does not exist and T is conservative and ergodic.

The reverse implication follows from the fact that for a non-negative measurable function f with $\int f \, d\mu > 0$ there must exist $\delta > 0$ with $\mu(\{f > \delta\}) > 0$. Hence, $\delta \mathbb{1}_{\{f>\delta\}} \leq f$ and by Lemma 2.4.3 and Remark 2.2.13 we find μ-a.e. that

$$\sum_{k=0}^{\infty} f \circ T^k \geq \delta \sum_{k=0}^{\infty} \mathbb{1}_{\{f>\delta\}} \circ T^k = \infty.$$

$\qquad\square$

Corollary 2.4.5. *If $(X, \mathcal{B}, \mu, T)$ is a non-singular system, then it is also conservative and ergodic if and only if for every measurable function $f \in L_1^+(\mu)$ with $\int f \, d\mu > 0$ we have that*

$$\sum_{k=0}^{\infty} \widehat{T}^k f = \infty, \quad \mu\text{-a.e. on } X.$$

Proof. Suppose first that T is conservative and ergodic. Since by Lemma 2.4.3 any $A \in \mathcal{B}$ with $\mu(A) > 0$ is a sweep-out set, we have that

$$\int \sum_{k=0}^{\infty} \widehat{T}^k f \cdot \mathbb{1}_A \, d\mu = \int f \cdot \sum_{k=0}^{\infty} \mathbb{1}_A \circ T^k \, d\mu = \infty.$$

Therefore, it follows that $\sum_{k=0}^{\infty} \widehat{T}^k f = \infty$ μ-a.e. on X.

For the reverse implication fix a wandering set $W \in \mathcal{B}$ with $\mu(W) > 0$ and $f \in L_1^+(\mu)$ with $\int f \, d\mu > 0$. Then we obtain the contradiction

$$\infty = \int \sum_{k=0}^{\infty} \widehat{T}^k f \cdot \mathbb{1}_W \, d\mu = \int f \cdot \sum_{k=0}^{\infty} \mathbb{1}_W \circ T^k \, d\mu \leq \int f \, d\mu < \infty.$$

If A is T-invariant with $\mu(A) > 0$ and B a measurable subset of $X \setminus A$ with $0 < \mu(B) < \infty$ then similarly we find the contradiction

$$\infty = \int \sum_{k=0}^{\infty} \widehat{T}^k \mathbb{1}_B \cdot \mathbb{1}_A \, d\mu = \int \mathbb{1}_B \cdot \sum_{k=0}^{\infty} \mathbb{1}_A \circ T^k \, d\mu = \int \sum_{k=0}^{\infty} \mathbb{1}_B \cdot \mathbb{1}_A \, d\mu = 0.$$

This shows that T is conservative and ergodic. $\qquad\square$

Proposition 2.4.6. *For a non-singular dynamical system $(X, \mathcal{B}, \mu, T)$, the following are equivalent:*
(a) *T is ergodic with respect to μ.*
(b) *For $B \in \mathcal{B}$, if $B = T^{-1}B \mod \mu$ (that is $\mu(B \triangle T^{-1}(B)) = 0$), then either $\mu(B) = 0$ or $\mu(X/B) = 0$.*
(c) *For $f : X \to \mathbb{R}$ measurable, if $f \circ T = f$ μ-a.e., then f is μ-a.e. equal to a constant.*

Proof. First we prove that (a) implies (b). Suppose that T is ergodic and let B be a μ-almost-invariant measurable set, that is, $\mu(B \triangle T^{-1}(B)) = 0$. We aim to construct a T-invariant set A from B such that A has the same μ-measure as B. So, define

$$A := \bigcap_{n=0}^{\infty} \bigcup_{k=n}^{\infty} T^{-k}(B).$$

It then follows, for each $n \geq 1$, that

$$B \triangle \bigcup_{k=n}^{\infty} T^{-k}(B) \subset \bigcup_{k=n}^{\infty} B \triangle T^{-k}(B).$$

Moreover, since

$$B \triangle T^{-k}(B) \subset \bigcup_{i=0}^{k-1} T^{-i}(B) \triangle T^{-(i+1)}(B) = \bigcup_{i=0}^{k-1} T^{-i}(B \triangle T^{-1}(B))$$

and since the system is non-singular, we also have that $\mu(B \triangle T^{-k}(B)) = 0$. Let $B_n := \bigcup_{k=n}^{\infty} T^{-k}(B)$ and notice that the sequence $(B_n)_{n \geq 1}$ is a decreasing sequence of sets with the property that $\mu(B_n \triangle B) = 0$, for each $n \in \mathbb{N}$, and $\bigcap_{n \in \mathbb{N}} B_n = A$. It follows that $\mu(A \triangle B) = 0$ and so,

$$\mu(A) = \mu(B).$$

Furthermore, it is clear that the set A is T-invariant. Hence, the ergodicity of T implies that $\mu(A) = 0$ or $\mu(X \setminus A) = 0$. Consequently, either $\mu(B) = 0$ or $\mu(X/B) = 0$, which proves the first implication.

Now we prove the implication from (b) to (c). Let the system be ergodic and suppose that for the measurable function $f : X \to \mathbb{R}$ we have that $f \circ T = f$ μ-a.e. For each $c \in \mathbb{R}$, we make the observation that the set $D_c := \{f \leq c\} := \{x \in X : f(x) \leq c\}$ is

(μ-a.e.) T-invariant. Therefore, for each of these sets D_c, either $\mu(D_c) = 0$ or $D_c = X$, up to a set of μ-measure zero. Let $c_0 := \inf\{c \in \mathbb{R} : D_c = X \bmod \mu\}$. Then,

$$\{f = c_0\} = \bigcap_{n \geq 1}\left\{f \leq c_0 + \frac{1}{n}\right\} \setminus \bigcup_{n \geq 1}\left\{f \leq c_0 - \frac{1}{n}\right\}$$
$$= X \bmod \mu.$$

In other words, $f(x) = c_0$ for μ-a.e. $x \in X$.

To see that (c) implies (a), suppose that for all measurable functions $f : X \to \mathbb{R}$, we have that if $f \circ T = f$ μ-a.e., then f is μ-a.e. equal to a constant. Let $B \in \mathcal{B}$ be such that $T^{-1}(B) = B$. Then, $\mathbb{1}_B \circ T = \mathbb{1}_B$ everywhere on X and so, by assumption, $\mathbb{1}_B$ is constant μ-a.e. on X. Therefore, by the definition of $\mathbb{1}_B$, we have that either $\mu(B) = 0$ or $\mu(X/B) = 0$. □

If we additionally assume that the system is conservative we can amend the list of equivalences in the following way.

Proposition 2.4.7. *For a conservative non-singular dynamical system $(X, \mathcal{B}, \mu, T)$, the following are equivalent:*
(a) *T is ergodic with respect to μ.*
(b) *For $A \in \mathcal{B}$, if $\mu(A) > 0$, then $\bigcup_{n=0}^{\infty} T^{-n}(A) = X \bmod \mu$.*
(c) *For $A, B \in \mathcal{B}$, if $\mu(A)\mu(B) > 0$, then there exists $n \in \mathbb{N}$ such that*

$$\mu(T^{-n}(A) \cap B) > 0.$$

Proof. We will prove the string of implications (a) implies (b) implies (c) implies (a).

The implication from (a) to (b) is a consequence of Lemma 2.4.3 since the system is supposed to be non-singular, ergodic and conservative. To prove that (b) implies (c), let A and B be sets of positive measure. Since (b) holds, we have that

$$\bigcup_{n=1}^{\infty} T^{-n}(A) = X \bmod \mu,$$

which implies that

$$0 < \mu(B) = \mu\left(\bigcup_{n=1}^{\infty} B \cap T^{-n}(A)\right) \leq \sum_{n=1}^{\infty} \mu(B \cap T^{-n}(A)).$$

It follows that there must be at least one $n \geq 1$ such that $\mu(B \cap T^{-n}(A)) > 0$.

Suppose now that (c) holds and let A be a T-invariant set. Then we have for all $n \in \mathbb{N}$ that

$$0 = \mu((X \setminus A) \cap A) = \mu((X \setminus A) \cap T^{-n}(A)).$$

So, by (c), either $\mu(A) = 0$ or $\mu(X \setminus A) = 0$, proving that T is ergodic. This finishes the string of equivalences. $\qquad\square$

Next we state the following important uniqueness result for finite invariant ergodic measures.

Proposition 2.4.8. *Let* $(X, \mathcal{B}, \mu, T)$ *be an ergodic invariant system with* $\mu(X) = 1$ *and let* ν *be another T-invariant probability measure on* $(X, \mathcal{B})$ *with* $\nu \ll \mu$. *Then we have that* $\nu = \mu$.

Proof. Since $\nu \ll \mu$, the Radon–Nikodým Theorem implies that there exists a density $f \in L_1^+(\mu)$ with $d\nu = f d\mu$. We are going to prove that f is constant μ-almost everywhere. Since ν and μ are both probability measures, this then guarantees that $f = 1$. In fact, for $r \in \mathbb{R}$ and all $B \subset \{f > r\}$ with positive μ-measure we have

$$\nu(B) - r\mu(B) = \int_B (f(x) - r)\, d\mu(x) > 0,$$

which implies that

$$\nu(B) > r\mu(B).$$

Similarly, for all $C \subset F_r := \{f \le r\}$ it follows that $\nu(C) \le r\mu(C)$. Making use of Remark 2.1.2 and since $T^{-1}(F_r) \setminus F_r \subset \{f > r\}$, we either have $\mu\left(T^{-1}(F_r) \setminus F_r\right) = 0$ or

$$\nu(T^{-1}(F_r) \setminus F_r) > r\mu(T^{-1}(F_r) \setminus F_r) = r\mu(F_r \setminus T^{-1}(F_r))$$
$$\ge \nu(F_r \setminus T^{-1}(F_r)) = \nu(T^{-1}(F_r) \setminus F_r),$$

which is impossible. Again in light of Remark 2.1.2, we have

$$\mu(T^{-1}(F_r) \triangle F_r) = \mu(T^{-1}(F_r) \setminus F_r) + \mu(F_r \setminus T^{-1}(F_r)) = 0.$$

Since T is ergodic, it follows that $\mu(F_r) \in \{0, 1\}$. This shows that f has to be μ-a.e. equal to a constant. $\qquad\square$

2.4.1 Ergodicity of the systems G and L_α

Let us now prove that the Gauss map is ergodic with respect to the Gauss measure. Before beginning, we introduce the notation "$a \asymp b$", which means that there exists a positive constant C such that $C^{-1}a \le b \le Ca$.

Lemma 2.4.9. *The Gauss map G is ergodic with respect to the Gauss measure m_G.*

Proof. The first and most important step in the proof is to show that for any given continued fraction cylinder set $C(x_1, \ldots, x_n)$ and for all measurable sets B, we have that

$$m_G(G^{-n}(B) \cap C(x_1, \ldots, x_n)) \asymp m_G(B)m_G(C(x_1, \ldots, x_n)). \tag{2.8}$$

In fact, it is sufficient to demonstrate (2.8) for all intervals of the form $B := [c, d] \subseteq [0, 1]$, since the set of all Borel sets satisfying (2.8) for a fixed constant can be shown to be a monotone class.

Thus, let B be some fixed interval in $[0, 1]$ and let $p_n/q_n := [x_1, \ldots, x_{n-1}, x_n]$ and $p_{n-1}/q_{n-1} := [x_1, \ldots, x_{n-1}]$ denote the n-th and $(n-1)$-th approximants of the number $x = [x_1, x_2, \ldots] \in [0, 1]$. Notice that $x \in G^{-n}(B) \cap C(x_1, \ldots, x_n)$ if and only if $G^n(x) = [x_{n+1}, x_{n+2}, \ldots] \in B$. Since G^n is monotonic on each cylinder set $C(x_1, \ldots, x_n)$, it follows that $G^{-n}(B) \cap C(x_1, \ldots, x_n)$ is an interval with endpoints given by

$$\frac{p_n + p_{n-1}c}{q_n + q_{n-1}c} \quad \text{and} \quad \frac{p_n + p_{n-1}d}{q_n + q_{n-1}d},$$

for some $c, d > 0$. (This follows from Theorem 1.1.5 and the fact that in this case we have $r_{n+1} = c^{-1}$ or $r_{n+1} = d^{-1}$.) Therefore, the Lebesgue measure of the intersection $G^{-n}(B) \cap C(x_1, \ldots, x_n)$ is equal to

$$\left| \frac{p_n + p_{n-1}d}{q_n + q_{n-1}d} - \frac{p_n + p_{n-1}c}{q_n + q_{n-1}c} \right| = \left| \frac{p_n q_{n-1}d + p_{n-1}q_n c - p_n q_{n-1}d - p_{n-1}q_n c}{(q_n + q_{n-1}c)(q_n + q_{n-1}d)} \right|$$

$$= |d - c| \frac{1}{(q_n + q_{n-1}c)(q_n + q_{n-1}d)},$$

by Theorem 1.1.1 (c). On the other hand, recall that the Lebesgue measure of the cylinder set $C(x_1, \ldots, x_n)$ is given by

$$\lambda(C(x_1, \ldots, x_n)) = \frac{1}{q_n(q_n + q_{n-1})},$$

which implies that

$$\lambda(G^{-n}(B) \cap C(x_1, \ldots, x_n)) = \lambda(B)\lambda(C(x_1, \ldots, x_n)) \frac{q_n(q_n + q_{n-1})}{(q_n + q_{n-1}c)(q_n + q_{n-1}d)}$$

$$\asymp \lambda(B)\lambda(C(x_1, \ldots, x_n)).$$

In light of Proposition 2.1.12, the proof of (2.8) is finished.

Now, suppose that $A \in \mathcal{B}$ is such that $G^{-1}(A) = A$. Then (2.8) implies that for each $n \in \mathbb{N}$ and for every cylinder set of level n we have that

$$m_G(A \cap C(x_1, \ldots, x_n)) \asymp m_G(A)m_G(C(x_1, \ldots, x_n)).$$

Clearly, this also holds for finite unions of (disjoint) cylinder sets and, since finite unions of cylinder sets generate the Borel σ-algebra, this implies that

$$m_G(A \cap B) \asymp m_G(A)m_G(B), \quad \text{for all } B \in \mathcal{B}.$$

On choosing $B := [0, 1] \setminus A$, it now follows that

$$0 \asymp m_G(A) m_G([0, 1] \setminus A),$$

which shows that $m_G(A) \in \{0, 1\}$. This finishes the proof. □

We can almost immediately obtain a stronger result about the Gauss map by only slightly altering the above proof. We aim to show that the Gauss map is an exact transformation. We first give the definition of exactness, for which we recall from Definition 2.2.4 that a transformation is said to be non-singular if it preserves sets of measure zero.

Definition 2.4.10. A non-singular transformation T of a σ-finite measure space $(X, \mathcal{B}, \mu)$ is said to be *exact* if for each B in the *tail σ-algebra* $\bigcap_{n \in \mathbb{N}} T^{-n}(\mathcal{B})$ we have that either $\mu(B)$ or $\mu(X \setminus B)$ is equal to zero.

Remark 2.4.11.
1. This definition of exactness only makes sense for non-invertible transformations. Indeed, if $T : X \to X$ is invertible, then it follows immediately that $T^{-n}(\mathcal{B}) = \mathcal{B}$ for every $n \in \mathbb{N}$. The correct corresponding property for invertible systems is the *K-property*, named for Kolmogorov, who introduced it. For more details, see [Par81] and references therein.
2. The tail σ-algebra is not an immediately transparent object. It helps to remember that it is an intersection of sets of sets. In particular, this means that if $B \in \bigcap_{n \in \mathbb{N}} T^{-n}(\mathcal{B})$, then $B \in T^{-n}(\mathcal{B})$ for all $n \in \mathbb{N}$. Thus, there exists a sequence of sets $(B_1, B_2, B_3, \ldots)$ such that $B = T^{-n}(B_n)$ for every $n \in \mathbb{N}$. Another way of thinking of this is to note that $T^{-n}(T^n(B)) = T^{-n}(T^n(T^{-n}(B_n))) = T^{-n}(B_n) = B$, for all $n \in \mathbb{N}$.
3. It is easy to see that an exact transformation must be ergodic, for if the map $T : X \to X$ is exact and B is a measurable subset of X such that $T^{-1}(B) = B$, then $T^{-n}(B) = B$ for all $n \in \mathbb{N}$ and so, the set B belongs to the tail σ-algebra and hence, either $\mu(B)$ or $\mu(X \setminus B)$ is equal to zero.

Theorem 2.4.12. *The Gauss map G is exact with respect to the Gauss measure m_G.*

Proof. Let $B \in \mathcal{B}$ be such that B lies in the tail σ-algebra $\bigcap_{n \in \mathbb{N}} G^{-n}(\mathcal{B})$. As noted in Remark 2.4.11, this implies that there exists a sequence of sets $(B_n)_{n \geq 1}$ such that $B = G^{-n}(B_n)$ for every $n \in \mathbb{N}$. We have shown in the proof of Lemma 2.4.9, in (2.8), that for cylinder sets $C(x_1, \ldots, x_n)$ we have

$$m_G(G^{-n}(A) \cap C(x_1, \ldots, x_n)) \asymp m_G(A) m_G(C(x_1, \ldots, x_n)), \quad \text{for all } A \in \mathcal{B}.$$

In particular, for our sequence $(B_n)_{n \geq 1}$ we then have that

$$m_G(G^{-n}(B_n) \cap C(x_1, \ldots, x_n)) \asymp m_G(B_n) m_G(C(x_1, \ldots, x_n)), \quad \text{for all } n \in \mathbb{N}.$$

But this implies, since $m_G(B_n) = m_G(G^{-n}(B_n))$ and the measure m_G is G-invariant, that for every cylinder set $C(x_1, \ldots, x_n)$, we have that

$$m_G(B \cap C(x_1, \ldots, x_n)) \asymp m_G(B) m_G(C(x_1, \ldots, x_n)).$$

This also holds for finite unions of cylinder sets and thus, since the cylinder sets generate the Borel σ-algebra, we deduce that

$$m_G(B \cap A) \asymp m_G(B) m_G(A), \quad \text{for all } A \in \mathcal{B}.$$

Choosing A to be the complement of B yields the desired result. □

In Section 2.5 we will prove that the Farey map is exact. One ingredient for the proof there will be the following lemma, the proof of which is again based on the relation in (2.8).

Lemma 2.4.13. *Let $\widehat{C}$ denote an arbitrary level $n + 1$ Farey cylinder set with code ending in the symbol 1. For each $B \in \mathcal{B}$ such that $B \subseteq \widehat{C}$, we then have that*

$$\lambda(B) \asymp \lambda(\widehat{C}) \lambda \left(F^n(B) \right).$$

Proof. Say that for some $x_1, \ldots, x_k \in \mathbb{N}$ with $\sum_{i=1}^k x_i = n + 1$ we have

$$\widehat{C} := \widehat{C} \left(0^{x_1-1}, 1, 0^{x_2-1}, \ldots, 1, 0^{x_k-1}, 1 \right).$$

Next consider some set $B \in \mathcal{B}$ such that $B \subseteq \widehat{C}$. Then there exists $E \in \mathcal{B}$ such that

$$B = G^{-k}(E) \cap C(x_1, \ldots, x_k) = F^{-(n+1)}(E) \cap \widehat{C}(0^{x_1-1}, 1, 0^{x_2-1}, \ldots, 1, 0^{x_k-1}, 1).$$

Since $E = G^k(B) = F^{n+1}(B)$ and since $\lambda \asymp m_G$, by Proposition 2.1.12, it follows from (2.8) that

$$\lambda(B) \asymp m_G(B) = m_G(G^{-k}(E) \cap C(x_1, \ldots, x_k)) \asymp m_G(E) m_G(C(x_1, \ldots, x_k))$$
$$\asymp \lambda(F^{n+1}(B)) \lambda(\widehat{C}) \asymp \lambda(F^n(B)) \lambda(\widehat{C}).$$

Here the final inequality follows by using the change of variables formula. □

Let us now prove directly that an α-Lüroth system $([0, 1], L_\alpha)$ is exact with respect to the Lebesgue measure. The proof follows along the same lines as that for the Gauss map, so we give only a sketch here and leave the details as an exercise for the reader.

Proposition 2.4.14. *The α-Lüroth map L_α is exact with respect to λ.*

Proof. To start, let $B \in \bigcap_{n \in \mathbb{N}} L_\alpha^{-n}(\mathcal{B})$ be given. We aim to prove that $\lambda(B) \in \{0, 1\}$. First, it is straightforward to calculate that for any single cylinder set $C_\alpha(\ell_1, \ldots, \ell_n)$, we have that

$$\lambda(B \cap C_\alpha(\ell_1, \ldots, \ell_n)) = \lambda(C_\alpha(\ell_1, \ldots, \ell_n))\lambda(B).$$

One immediately verifies that this also holds for a finite union of L_α-cylinder sets. From this, we deduce that

$$\lambda(B \cap C) = \lambda(B)\lambda(C), \text{ for all } C \in \mathcal{B}.$$

Therefore, by choosing C to be equal to $[0, 1] \setminus B$, we conclude that

$$0 = \lambda(B \cap ([0, 1] \setminus B)) = \lambda(B)\lambda([0, 1] \setminus B).$$

This shows that $\lambda(B) = 0$ or $\lambda(B) = 1$, and hence finishes the proof. □

Corollary 2.4.15. *The α-Lüroth map L_α is ergodic with respect to λ.*

2.4.2 Ergodic theorems for probability spaces and consequences for the Gauss and α-Lüroth systems

The first major result in ergodic theory was published in 1931 by G.D. Birkhoff [Bir31][1]. This result is known as the pointwise ergodic theorem and it gives a precise relationship between the average of an integrable function evaluated along the orbit of a typical point (the time average) and the integral of the function (the space average). There are now a great variety of proofs available; the interested reader is referred to either [Wal82] or [EW11] (and references therein). In Chapter 4 we will prove the more general Chacon–Ornstein Ergodic Theorem and then show how to derive Birkhoff's Pointwise Ergodic Theorem from this more general theorem. Let us here simply state the result for the case of an ergodic probability-measure-preserving system.

Theorem 2.4.16 (Birkhoff's Pointwise Ergodic Theorem). *Let $(X, \mathcal{B}, \mu, T)$ be an ergodic, probability-measure-preserving system. If $f \in L_1(\mu)$, then we have for μ-a.e. $x \in X$ that*

$$\lim_{n \to \infty} \frac{1}{n} \sum_{j=0}^{n-1} f \circ T^j(x) = \int f \, d\mu.$$

---

[1] Birkhoff's Pointwise Ergodic Theorem, although published first, was not the first ergodic theorem to be proved. The work of von Neumann [Neu32] predates that of Birkhoff. In [Neu32] von Neumann proves what is now called the Mean Ergodic Theorem. See the book [EW11] for an exposition of this result and further references.

Since we have already proved that the Gauss and α-Lüroth systems are ergodic and probability-measure-preserving, we may apply Birkhoff's Pointwise Ergodic Theorem to easily obtain the following interesting number-theoretic results. The original proofs (in the continued fraction case, that is) of most of these statements were decidedly more complicated.

Proposition 2.4.17. *For λ-almost every real number $x = [x_1, x_2, x_3, \ldots] \in [0,1]$, the following statements hold:*

(a) *The element j appears in the continued fraction expansion of x with frequency*

$$\lim_{n\to\infty} \frac{1}{n} \#\{i : i \le n, x_i = j\} = \frac{2\log(1+j) - \log(j) - \log(2+j)}{\log 2}.$$

More generally, every finite sequence $y_1, \ldots, y_n$ of positive integers appears in the continued fraction expansion of x with frequency

$$m_G(C(y_1, \ldots, y_n)).$$

(b) *For the geometric mean of the elements x_n we have*

$$\lim_{n\to\infty} (x_1 x_2 \ldots x_n)^{1/n} = \prod_{n=1}^{\infty} \left(\frac{(n+1)^2}{n(n+2)} \right)^{\log n / \log 2}.$$

(c) *For the arithmetic mean of the elements x_n, we have*

$$\lim_{n\to\infty} \frac{1}{n}(x_1 + x_2 + \cdots + x_n) = \infty.$$

(d) *For the growth rate of the denominators q_n of the approximants to x, we have*

$$\lim_{n\to\infty} \frac{1}{n} \log(q_n) = \frac{\pi^2}{12\log 2}.$$

(e) *For the rate at which the approximants p_n/q_n converge to x, we have*

$$\lim_{n\to\infty} \frac{1}{n} \log \left| x - \frac{p_n}{q_n} \right| = -\frac{\pi^2}{6\log 2}.$$

Proof. For the first statement, first notice that the element j appears in the first n elements of the continued fraction expansion of an irrational number x with frequency

$$\frac{1}{n} \#\{i : i \le n, x_i = j\} = \frac{1}{n} \# \left\{ i : i \le n, G^i(x) \in \left(\frac{1}{j+1}, \frac{1}{j} \right) \right\}.$$

Now, observe that if we define the function $f \in L_1(m_G)$ to be equal to the characteristic function $f := \mathbb{1}_{(1/(j+1),1/j)}$, then the ergodic sum $\frac{1}{n} \sum_{j=0}^{n-1} f \circ G^j(x)$ coincides with the n-th frequency defined above. Thus, by Birkhoff's Pointwise Ergodic Theorem, we immediately deduce that

$$\lim_{n\to\infty}\frac{1}{n}\#\{i:i\le n,x_i=j\}=\int f\,dm_G=\frac{1}{\log 2}\int_{1/(j+1)}^{1/j}\frac{1}{1+y}\,d\lambda(y)$$

$$=\frac{1}{\log 2}\left(\log\left(1+\frac{1}{j}\right)-\log\left(1+\frac{1}{j+1}\right)\right)$$

$$=\frac{2\log(1+j)-\log(j)-\log(2+j)}{\log 2}.$$

The proof of the remaining part of the first statement follows similarly, on choosing $f:=\mathbb{1}_{C(y_1,\dots,y_n)}$.

For part (b), define the function $f:(0,1)\to(0,1)$ by setting $f(x):=\log n$, for $x\in(1/(n+1),1/n)$. It is easy to check that the function f is in $L_1(\lambda)$ (and hence in $L_1(m_G)$, since λ and m_G are comparable). By Birkhoff's Pointwise Ergodic Theorem, we therefore have for λ-a.e. x that

$$\lim_{n\to\infty}\frac{1}{n}\sum_{j=1}^{n}\log x_j=\lim_{n\to\infty}\frac{1}{n}\sum_{j=0}^{n-1}f(G^j(x))=\int f\,dm_G.$$

A simple calculation shows that this yields the identity in part (b).

Proving part (c) requires a little more effort. Let now the function f be defined by $f(x):=\lfloor 1/x\rfloor=x_1$, that is, $f(x)$ is defined to be equal to the first element in the continued fraction expansion of x. We then have that

$$\frac{1}{n}(x_1+x_2+\cdots+x_n)=\frac{1}{n}\sum_{j=0}^{n-1}f(G^j(x)).$$

However, we cannot directly apply Birkhoff's Pointwise Ergodic Theorem because the function f is not integrable in this instance. To overcome this, define for each $N\in\mathbb{N}$,

$$f_N(x):=\begin{cases}f(x)&\text{if }f(x)\le N;\\0&\text{otherwise.}\end{cases}$$

The function f_N is in $L_1(\lambda)$ and so, by Birkhoff's Pointwise Ergodic Theorem, we have that

$$\liminf_{n\to\infty}\frac{1}{n}\sum_{j=0}^{n-1}f(G^j(x))\ge\lim_{n\to\infty}\frac{1}{n}\sum_{j=0}^{n-1}f_N(G^j(x))$$

$$=\int_0^1 f_N\,dm_G.$$

The fact that the above integral tends to infinity as N tends to infinity finishes the proof of part (c).

In order to prove part (d), first observe that if $x = [x_1, x_2, x_3, \ldots]$, then

$$\frac{p_n(x)}{q_n(x)} = \frac{1}{x_1 + [x_2, \ldots, x_n]} = \frac{1}{x_1 + \dfrac{p_{n-1}(G(x))}{q_{n-1}(G(x))}}$$

$$= \frac{q_{n-1}(G(x))}{x_1 q_{n-1}(G(x)) + p_{n-1}(G(x))}.$$

This shows that $p_n(x) = q_{n-1}(G(x))$ for every $n \in \mathbb{N}$, since the approximants are in reduced form. It follows that

$$\frac{1}{q_n(x)} = \frac{p_n(x)}{q_n(x)} \cdot \frac{p_{n-1}(G(x))}{q_{n-1}(G(x))} \cdots \frac{p_1(G^{n-1}(x))}{q_1(G^{n-1}(x))},$$

so that

$$-\frac{1}{n} \log(q_n(x)) = \frac{1}{n} \sum_{j=0}^{n-1} \log\left(\frac{p_{n-j}(G^j(x))}{q_{n-j}(G^j(x))}\right).$$

Let the $L_1(m_G)$-function f be defined by $f(x) := \log x$. It then follows that

$$-\frac{1}{n} \log(q_n(x)) = \frac{1}{n} \sum_{j=0}^{n-1} f(G^j(x)) - \frac{1}{n} \sum_{j=0}^{n-1} \left(f(G^j(x)) - f\left(\frac{p_{n-j}(G^j(x))}{q_{n-j}(G^j(x))}\right)\right).$$

First noticing that the second term on the right hand side tends to zero as n tends to infinity, since $\frac{p_{n-j}(G^j(x))}{q_{n-j}(G^j(x))}$ is a good approximation to $G^j(x)$ for large n, we have by Birkhoff's Pointwise Ergodic Theorem that

$$\lim_{n \to \infty} -\frac{1}{n} \log q_n = \lim_{n \to \infty} \frac{1}{n} \sum_{j=0}^{n-1} f(G^j(x)) = \frac{1}{\log 2} \int_0^1 \frac{\log x}{1+x} \, d\lambda(x) = -\frac{\pi^2}{12 \log 2}.$$

This proves part (d). Finally, since

$$\log q_n + \log q_{n+1} \leq -\log\left|x - \frac{p_n}{q_n}\right| \leq \log q_n + \log q_{n+2},$$

part (e) follows from part (d). This finishes the proof of the proposition. □

Remark 2.4.18. Part (a) of the above proposition implies that for λ-a.e. $x \in [0, 1]$, the continued fraction expansion of x contains two 1s in a row infinitely often. Notice that this implies that the same property holds for the Farey coding of almost every point. This will turn out to be useful later on, when we prove that the Farey map is exact.

Part (c) says that the arithmetic mean of the first n continued fraction digits diverges a.e. as n tends to infinity. Nevertheless, there exist meaningful stochastic laws describing the continued fraction digits in greater depth. Lévy [Lév52] was the first to derive non-degenerate limit laws in the context of continued fractions namely, we have that the continued fraction digits belong to the domain of attraction to a stable law with characteristic exponent 1. More precisely we have the following convergence in distribution with respect to any absolutely continuous probability measure $\mu \ll \lambda$

$$\frac{S_k}{k/\log 2} - \log k \xrightarrow{\mu} F,$$

where F has a *stable distribution* (cf. [Hei87] and [Phi88], and for related results see also [Hen00]).

Khinchin showed that for a suitable normalising sequence a weak law of large numbers holds (cf. [Khi35]). That is,

$$\frac{S_n}{(n \log n)} \to \frac{1}{\log 2}$$

converges in measure with respect to λ. However, according to [Phi88] there is no normalising sequence (n_k) with (n_k/k) non-decreasing such that a strong law of large numbers is satisfied. More precisely, we either have that

$$\sum_{k=1}^{\infty} \frac{1}{n_k} < \infty \text{ and } \lim_{k \to \infty} \frac{S_k}{n_k} = 0 \; \lambda\text{-a.e.}$$

or

$$\sum_{k=1}^{\infty} \frac{1}{n_k} = \infty \text{ and } \lim_{k \to \infty} \frac{S_k}{n_k} = \infty \; \lambda\text{-a.e.}$$

On the other hand, Diamond and Vaaler have shown in [DV86] that for the trimmed sum

$$S_n^\flat := \sum_{i=1}^{n} x_i - \max_{1 \leq \ell \leq n} x_\ell$$

we have

$$\lim_{n \to \infty} \frac{S_n^\flat}{n \log n} = \frac{1}{\log 2} \; \lambda\text{-a.e.}$$

Finally, let us mention two further related results, namely, the extreme value law for continued fractions by Galambos [Gal73, JKS13] and Philipp's law [Phi76]. The extreme value law states

$$\lim_{n \to \infty} \lambda \left\{ \max_{1 \leq k \leq n} x_k < \frac{ns}{\log(2)} \right\} = \exp(-1/s)$$

and Philipp showed that Lebesgue a.e. we have

$$\liminf_{n \to \infty} \max_{1 \leq k \leq n} x_k \cdot \frac{\log \log n}{n} = \frac{1}{\log 2}.$$

Let us now turn our attention to the α-Lüroth systems. In light of the fact that we proved in Section 2.4.1 that each map L_α is ergodic with respect to the Lebesgue measure, we can use Birkhoff's Pointwise Ergodic Theorem to obtain various statements about the α-Lüroth elements of λ-a.e. real number $x \in [0, 1]$.

Proposition 2.4.19. *Let L_α denote the α-Lüroth map for the partition $\alpha = \{A_n : n \in \mathbb{N}\}$ with $\lambda(A_n) = a_n$ and tails $t_n = \sum_{k=n}^\infty a_k$, as before. Then, for λ-a.e. $x = [\ell_1, \ell_2, \ell_3, \ldots]_\alpha \in [0, 1]$, the following statements hold:*

(a) $\displaystyle \lim_{n\to\infty} \frac{1}{n} \#\{j \le n : \ell_j = k\} = a_k$, *for each $k \in \mathbb{N}$.*

(b) $\displaystyle \lim_{n\to\infty} \frac{1}{n} \log \left(\prod_{j=1}^n \ell_j \right) = \sum_{k=1}^\infty a_k \log k.$

(c) $\displaystyle \lim_{n\to\infty} \frac{1}{n} \sum_{j=1}^n \ell_j = \sum_{k=1}^\infty t_k.$

(d) *For each $k \in \mathbb{N}$, every finite sequence $y_1, \ldots, y_k$ of positive integers appears infinitely often in the α-Lüroth expansion of x.*

(e) *With the additional assumption on the partition α that $a_n \le t_{n+1}$ for sufficiently large $n \in \mathbb{N}$, we have that*

$$\lim_{n\to\infty} \frac{1}{n} \log \left| x - r_n^{(\alpha)} \right| = \sum_{k=1}^\infty a_k \log a_k.$$

Proof. Each of the above statements follows directly on application of Birkhoff's Pointwise Ergodic Theorem to a specific Lebesgue integrable function f. For the first four assertions, choose the function f in turn to be given by the characteristic function $\mathbb{1}_{A_k}$, then $\log(\ell_1(x))$, then $\ell_1(x)$ and finally by the characteristic function $\mathbb{1}_{C_\alpha(y_1,\ldots,y_k)}$. Here, $\ell_1(x)$ is defined to be the first element in the α-Lüroth expansion of x. For part (c), note that we have to do the same trick as was done for the Gauss map case only in the event that the partition α is of infinite type.

For part (e), first notice that under the stated condition on α, we have for sufficiently large n that

$$a_{\ell_1} \ldots a_{\ell_n} a_{\ell_{n+1}} \le \left| x - r_n^{(\alpha)} \right| \le a_{\ell_1} \ldots a_{\ell_n}. \tag{2.9}$$

Then, let f be given by $f(x) := \log(a_{\ell_1(x)})$. Using (2.9), we have that

$$\lim_{n\to\infty} \frac{1}{n} \sum_{j=0}^{n-1} f \circ L_\alpha^j(x) = \lim_{n\to\infty} \frac{1}{n} \sum_{j=1}^n \log a_{\ell_j(x)} = \lim_{n\to\infty} \frac{1}{n} \sum_{j=1}^n \log \left| x - r_n^{(\alpha)} \right|$$

$$= \int_{[0,1]} \log a_{\ell_1(x)} \, d\lambda(x) = \sum_{k=1}^\infty \int_{A_k} \log a_{\ell_1(x)} \, d\lambda(x)$$

$$= \sum_{k=0}^\infty a_k \log a_k.$$

This finishes the proof. $\qquad\square$

Remark 2.4.20.
1. The lists given in Propositions 2.4.17 and 2.4.19 gives only a small sample of the possible results obtainable using Birkhoff's Pointwise Ergodic Theorem in this manner. The reader is invited to think of others.
2. It is immediately clear that the densities of the appearances of the digits in the α-Lüroth expansion constitute a probability vector, as they are just given by the associated a_ks. Calculating the sum of the frequencies appearing in part (a) of Proposition 2.4.17 shows that the same is true for the Gauss map.
3. The extra condition on α given in part (e) of the previous proposition is equivalent to the requirement that $a_n/t_n \leq 1/2$, for all n sufficiently large. For the example of the alternating Lüroth map, this condition is met. It is also satisfied for any expansive partition of exponent $\theta > 0$ and for expanding partitions with $\rho < 2$.

To finish this section we will prove a useful consequence of Birkhoff's Ergodic Theorem.

Proposition 2.4.21. *Under the assumptions of Birkhoff's Pointwise Ergodic Theorem we also have convergence of $\left(n^{-1}\sum_{j=0}^{n-1}f\circ T^j\right)$ in $L_1(\mu)$.*

Proof. To prove convergence in $L_1(\mu)$ we first verify the claim for bounded functions and then use the fact that $L_\infty(\mu)$ is dense in $L_1(\mu)$. Let $h \in L_\infty(\mu) \subset L_1(\mu)$. For notational convenience we write $S_n h := \sum_{j=0}^{n-1} h \circ T^j$. Since $\|h \circ T\|_\infty = \|h\|_\infty$ we also have that the a.e. defined limit $h^* = \lim n^{-1} S_n h$ is in $L_\infty(\mu)$. Hence, a.e. we have $|n^{-1}S_n h - h^*| \to 0$ and by Lebesgue's Dominated Convergence Theorem, it follows that $\|n^{-1}S_n h - h^*\|_1 \to 0$. Since $(n^{-1}S_n h)$ is a Cauchy sequence in the Banach space $L_1(\mu)$, for every $\varepsilon > 0$ there exists $N(\varepsilon, h) \in \mathbb{N}$, such that for all $k > 0$ and $n > N(\varepsilon, h)$ we have

$$\left\| \frac{1}{n}S_n h - \frac{1}{n+k}S_{n+k}h \right\|_1 < \varepsilon.$$

For $\varepsilon > 0$ and each $f \in L_1(\mu)$ we can find $h \in L_\infty(\mu)$ with $\|f - h\|_1 < \varepsilon/4$. Then for $n > N(\varepsilon/2, h)$ and $k > 0$

$$\left\| \frac{1}{n}S_n f - \frac{1}{n+k}S_{n+k}f \right\|_1 \leq \left\| \frac{1}{n}S_n f - \frac{1}{n}S_n h \right\|_1 + \left\| \frac{1}{n}S_n h - \frac{1}{n+k}S_{n+k}h \right\|_1$$
$$+ \left\| \frac{1}{n+k}S_{n+k}h - \frac{1}{n+k}S_{n+k}f \right\|_1$$
$$\leq 2\|f - h\|_1 + \varepsilon/2 < \varepsilon.$$

Therefore $\left(\frac{1}{n}S_n f\right)_{n\geq 1}$ is a Cauchy sequence in $L_1(\mu)$ and consequently must have a limit. This finishes the proof. □

2.4.3 Ergodic theorems for infinite measures

Let us now turn our attention back to infinite measure-preserving systems. It turns out, and we give a straightforward proof of this at the end of the section, that in the case of a dynamical system that preserves an infinite measure, Birkhoff's Pointwise Ergodic Theorem is replaced by the following statement.

Theorem 2.4.22. *Let (X, B, μ, T) be a conservative and ergodic measure-preserving system, such that $\mu(X) = \infty$. Then, for all $f \in L_1(\mu)$ and μ-a.e. $x \in X$, we have that*

$$\lim_{n \to \infty} \frac{1}{n} \sum_{j=0}^{n-1} f \circ T^j(x) = 0.$$

We delay the proof of the above theorem until after the statement of a stronger ergodic theorem (Theorem 2.4.24); for the moment, let us consider again the original statement of Birkhoff's Pointwise Ergodic Theorem given in Theorem 2.4.16. So, let $T : (X, B, \mu) \to (X, B, \mu)$ be a probability-measure-preserving system and let $A \in B$ be a measurable set. Define $S_n(A) := \sum_{j=0}^{n-1} \mathbb{1}_A \circ T^j$, that is, the function $S_n(A)$ evaluated at a point simply counts the number of visits the orbit of x makes to the set A before time n. We shall, following Zweimüller [Zwe04], call $S_n(A)$ the *occupation time* of A. Birkhoff's Pointwise Ergodic Theorem then implies that

$$\lim_{n \to \infty} \frac{1}{n} S_n(A)(x) = \mu(A), \text{ for } \mu\text{-a.e. } x \in X.$$

This tells us three things. Firstly, it shows that the rate at which the occupation time of A diverges is asymptotically the same for μ-a.e. $x \in X$. Secondly, it proves that this rate depends on A only through the measure of the set A and, thirdly, it identifies the occupation time as being proportional to n.

For infinite systems, however, the infinite-measure version of Birkhoff's Pointwise Ergodic Theorem, given above in Theorem 2.4.22, only provides an upper bound for $S_n(A)$, but it gives no information on how the asymptotic behaviour of $S_n(A)(x)$ is related to A and to x. It is natural to ask whether a sequence $(c_n)_{n \in \mathbb{N}}$ of *normalising constants* can be found such that for all $A \in B$, we have that

$$\lim_{n \to \infty} \frac{1}{c_n} S_n(A)(x) = \mu(A), \text{ for } \mu\text{-a.e. } x \in X.$$

Unfortunately, this is simply not possible, as the next theorem shows.

Theorem 2.4.23 (Aaronson's Theorem). *Let $(X, \mathcal{B}, \mu, T)$ be a conservative and ergodic measure-preserving system, such that $\mu(X) = \infty$. Also let $(c_n)_{n \geq 1}$ be any arbitrary sequence of strictly positive real numbers. Then, for all $f \in L_1^+(\mu)$ we have,*

$$\liminf_{n \to \infty} \frac{1}{c_n} \sum_{j=0}^{n-1} f \circ T^j(x) = 0, \ \mu\text{-a.e.},$$

or there exists a sequence $(n_k) \in \mathbb{N}^{\mathbb{N}}$ with $n_k \to \infty$ such that for all $f \in L_1^+(\mu)$ with $\int f \, d\mu > 0$, we have,

$$\lim_{k \to \infty} \frac{1}{c_{n_k}} \sum_{j=0}^{n_k-1} f \circ T^j = \infty, \ \mu\text{-a.e.}$$

Proof. For the proof, we refer to Section 2.4 of Aaronson [Aar97]. □

Aaronson's result shows that in the situation of an infinite invariant measure, the asymptotic behaviour of ergodic sums (or, more specifically, occupation times) is extremely complicated. Despite this negative result by Aaronson, there are plenty of interesting qualitative and quantitative characterisations for infinite dynamical systems. Our first aim in this direction is to show that although the pointwise asymptotics of the ergodic sum of an integrable function f crucially depends on the point $x \in X$ chosen, it only depends upon the function f through its expected value $\int_X f \, d\mu$.

Theorem 2.4.24 (Hopf's Ratio Ergodic Theorem). *Let $(X, \mathcal{B}, \mu, T)$ be a conservative and ergodic measure-preserving system and let $f, g \in L_1(\mu)$, with $g \geq 0$ and $\int_X g \, d\mu > 0$. Then, for μ-a.e. $x \in X$, we have that*

$$\lim_{n \to \infty} \frac{\sum_{j=0}^{n-1} f \circ T^j(x)}{\sum_{j=0}^{n-1} g \circ T^j(x)} = \frac{\int_X f \, d\mu}{\int_X g \, d\mu}.$$

We will prove this theorem shortly, in Section 2.4.6, using the technique of *inducing*. (In Chapter 4 we will give another, alternative proof of this theorem, by showing how to derive it from the more general Chacon–Orstein Ergodic Theorem.) Before doing either of those things, let us now show how to prove Theorem 2.4.22 using Hopf's theorem: Assume that the measure μ is infinite. By the σ-finiteness of the space $(X, \mathcal{B}, \mu)$ we have that for each $m \in \mathbb{N}$, there exists some set $B_m \in \mathcal{B}$ such that $m \leq \mu(B_m) < \infty$. Applying Hopf's Ratio Ergodic Theorem to the functions $f \in L_1(\mu)$ and $\mathbb{1}_{B_m}$ yields that

$$0 \leq \limsup_{n \to \infty} \frac{1}{n} \sum_{j=0}^{n-1} f \circ T^j \leq \lim_{n \to \infty} \frac{\sum_{j=0}^{n-1} f \circ T^j}{S_n(\mathbb{1}_{B_m})} = \frac{\int_X f \, d\mu}{\mu(B_m)}, \ \mu\text{-a.e. on } X.$$

Here, the second inequality above comes from the fact that $S_n(\mathbb{1}_{B_m}) \leq n$. Since m was arbitrary and $\lim_{m \to \infty} \mu(B_m) = \infty$, the proof of Theorem 2.4.22 is complete.

2.4.4 Inducing

In this section we will introduce and study *induced maps*. The idea behind these maps is similar to that of the jump transformation introduced earlier (see Definitions 1.3.2 and 2.3.23). The basic construction goes back to Kakutani [Kak43] and Rokhlin [Roh48]. In essence, it consists of viewing an infinite measure-preserving system through the window of a set of finite measure. Recall that for a non-singular system $T : (X, \mathcal{B}, \mu) \to (X, \mathcal{B}, \mu)$, a set A is called a sweep-out set for T if $\bigcup_{n=0}^{\infty} T^{-n}(A) = X$ mod μ, and also that we showed in Lemma 2.4.3 that for conservative, ergodic transformations, every set $A \in \mathcal{B}$ with $0 < \mu(A) < \infty$ is a sweep-out set.

Definition 2.4.25. Let A be a sweep-out set for the non-singular, conservative transformation $T : X \to X$ and define the function $\varphi : X \to \mathbb{N}$ by setting

$$\varphi(x) := \inf\{n \geq 1 : T^n(x) \in A\}.$$

Note that the conservativity of T ensures that

$$A = A^* := A \cap \bigcap_{n \in \mathbb{N}} \bigcup_{k \geq n} T^{-k} A \quad \text{mod } \mu.$$

In the context of inducing we will always assume that $A = A^*$, which guarantees that $\varphi(x) < \infty$ for all $x \in A$. When we restrict the function φ to the set A, the map φ is called the *return time to A*. Finally, the *induced map* $T_A : A \to A$ of T on $A = A^*$ is defined to be

$$T_A(x) := T^{\varphi(x)}(x).$$

We refer to $(A, \mathcal{B}_A, m|_A, T_A)$ with $m|_A$ denoting the measure m restricted to $\mathcal{B}_A := \{A \cap B : B \in \mathcal{B}\}$ as the *induced system*. The idea behind T_A is that it is an accelerated version of T. It only records the times each orbit visits the set A, cutting out what happens in between. Let us now prove one straightforward, but nevertheless useful, identity. To shorten the notation, in all that follows we will again write $\{\varphi = n\}$ for the set $\{x \in X : \varphi(x) = n\}$ and, similarly, $\{\varphi > n\}$ for the set $\{x \in X : \varphi(x) > n\}$.

Lemma 2.4.26. *If A is a sweep-out set for the non-singular, conservative transformation $T : X \to X$, then we have for all $B \subseteq X$ that*

$$T_A^{-1}(A \cap B) = \bigcup_{n=1}^{\infty} A \cap \{\varphi = n\} \cap T^{-n}(B).$$

Further, the induced system $(A, \mathcal{B}_A, m|_A, T_A)$ is also non-singular and conservative.

Proof. The function φ is finite μ-almost everywhere on X. It therefore follows that $A = \bigcup_{n=1}^{\infty} A \cap \{\varphi = n\}$, where this union is disjoint. Since $T_A = T^n$ on the set $A \cap \{\varphi = n\}$, we have that

$$T_A^{-1}(A \cap B) = \bigcup_{n=1}^{\infty} A \cap \{\varphi = n\} \cap T_A^{-1}(A \cap B)$$

$$= \bigcup_{n=1}^{\infty} A \cap \{\varphi = n\} \cap T_A^{-1}(A) \cap T_A^{-1}(B)$$

$$= \bigcup_{n=1}^{\infty} A \cap \{\varphi = n\} \cap T^{-n}(B).$$

This identity immediately implies that the induced system is non-singular. To see that the induced system is also conservative note that by the definition of φ and the conservativity of T we have for all $B \in \mathcal{B}_A$ that

$$\sum_{k \in \mathbb{N}} \mathbb{1}_B \circ T_A^k = \sum_{n \in \mathbb{N}} \mathbb{1}_B \circ T^n = \infty$$

a.e. on B. □

Let us now turn to measure-theoretic questions. Properties of the induced map can be used to deduce interesting properties of the original system and vice versa, as we shall show in the following three propositions. First we assume some knowledge of T_A and use this to investigate T.

Proposition 2.4.27. *Let $T : X \to X$ be a $\mathcal{B}$-measurable transformation and assume that for $A \in \mathcal{B}$ the induced map $T_A : A \to A$ preserves some finite measure ν. Then the following hold:*

(a) *There exists a T-invariant measure m given by*

$$m(B) := \sum_{n=0}^{\infty} \nu(A \cap \{\varphi > n\} \cap T^{-n}(B)), \quad \text{for all } B \in \mathcal{B}$$

such that $m|_A = \nu$.

(b) *The system $(X, \mathcal{B}, m, T)$ is conservative.*

Proof. The proof of part (a) follows from Lemma 2.4.26, similarly to the proof of Lemma 2.3.25. We leave the details as an exercise. Part (b) follows directly from Maharam's Recurrence Theorem (see Theorem 2.2.14). □

Proposition 2.4.28. *Let $(X, \mathcal{B}, \mu, T)$ be conservative and non-singular.*

(a) *Assume that A is a sweep-out set for T such that the induced system $(A, \mathcal{B}_A, \mu|_A, T_A)$ is ergodic. Then $(X, \mathcal{B}, \mu, T)$ is also ergodic.*

(b) *If $(X, \mathcal{B}, \mu, T)$ is ergodic, then also $(A, \mathcal{B}_A, \mu|_A, T_A)$ is ergodic.*

Proof. To prove (a) we first claim that for any measurable T-invariant set $B = T^{-1}(B)$, the intersection $A \cap B$ is invariant under T_A. Indeed, in light of Lemma 2.4.26, we have that

$$T_A^{-1}(A \cap B) = \bigcup_{n=1}^{\infty} A \cap \{\varphi = n\} \cap T^{-n}(B) = \bigcup_{n=1}^{\infty} A \cap \{\varphi = n\} \cap B = A \cap B.$$

By assumption, T_A is ergodic, so either $\mu|_A(A \cap B) = 0$ or $\mu|_A((A \cap B)^c) = 0$. In the first case, we have $\mu(A \cap B) = 0$ and we may conclude, since A is a sweep-out set and T is non-singular, that

$$\mu(B) = \mu\left(\bigcup_{n \in \mathbb{N}} T^{-n}(A) \cap B\right) = \mu\left(\bigcup_{n \in \mathbb{N}} T^{-n}(A \cap B)\right) = 0.$$

Analogously, the second case yields that $m(B^c) = 0$ and the proof is finished.

For the proof of part (b) we make use of Lemma 2.4.3 in the following way. Assume that we have a set $B \in \mathcal{B}_A$ with $T_A^{-1}(B) = B$, $\mu_A(B) > 0$ and such that B is not equal to $A \mod \mu$. Then the set $A \setminus B$ has positive measure and for all $x \in A \setminus B$ we find the contradiction

$$0 = \sum_{k=0}^{\infty} \mathbb{1}_B \circ T_A^k(x) = \sum_{n=0}^{\infty} \mathbb{1}_B \circ T^n(x) = \infty.$$

□

Let us now present a converse to Proposition 2.4.28. That is, we now assume some knowledge concerning the original map T and use this knowledge to obtain facts about the induced map T_A.

Proposition 2.4.29. *Let $(X, \mathcal{B}, \mu, T)$ be a measure-preserving system and let A be a sweep-out set for T. Then the induced system $(A, \mathcal{B}_A, \mu|_A, T_A)$ is also measure-preserving.*

Proof. Fix $B \in \mathcal{B}_A$. Using Lemma 2.4.26 we find

$$\mu|_A\left(T_A^{-1}(B)\right)$$

$$= \sum_{n=1}^{\infty} \mu\left(A \cap \{\varphi = n\} \cap T_A^{-1}(B)\right)$$

$$= \sum_{n=1}^{\infty} \mu\left(T^{-1}\left(T^{-n+1}(B) \cap \bigcap_{k=0}^{n-2} T^{-k}(A^c)\right) \setminus \left(T^{-n}(B) \cap \bigcap_{k=0}^{n-1} T^{-k}(A^c)\right)\right)$$

$$= \sum_{n=1}^{\infty} \mu\left(T^{-n+1}(B) \cap \bigcap_{k=0}^{n-2} T^{-k}(A^c)\right) - \mu\left(T^{-n}(B) \cap \bigcap_{k=0}^{n-1} T^{-k}(A^c)\right)$$

$$= \mu|_A(B) - \lim_{n \to \infty} \mu\left(T^{-n}(B) \cap \bigcap_{k=0}^{n-1} T^{-k}(A^c)\right) \le \mu|_A(B).$$

On the other hand, applying the above observation to $A \setminus B \in \mathcal{B}_A$ instead of B, we infer that

$$\mu|_A \left(T_A^{-1}(B) \right) = \mu(A) - \mu(T_A^{-1}(A \setminus B)) \geq \mu(A) - \mu(A \setminus B) = \mu(B).$$

Combining these two inequalities proves the assertion. □

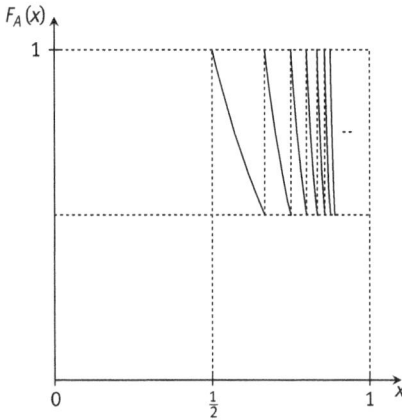

Fig. 2.1. The induced map F_A of the Farey map on the interval $A := [1/2, 1]$.

Example 2.4.30. For the Farey map F, let $A := [1/2, 1]$. Then, the induced map $F_A : A \to A$ is given by

$$F_A(x) := F^{\varphi(x)}(x) = F^{x_2(x)}(x), \quad \text{for } x = [1, x_2, x_3, \ldots].$$

So, in this case, the sets $\{\varphi = n\}$ are equal to the collection of second-level Gauss cylinder sets $\{C(1, n) : n \in \mathbb{N}\}$. In fact, we can explicitly calculate the map F_A as follows:

$$F_A(x) = \frac{1 - x}{nx - (n - 1)}, \quad \text{for } x = [1, n, x_3, x_4, \ldots] \in C(1, n).$$

Also, note that the action of F_A on the continued fraction expansion of a point $x = [1, x_2, x_3, \ldots] \in A$ is given by $F_A([1, x_2, x_3, x_4, \ldots]) = [1, x_3, x_4, \ldots]$. (You will be asked to check this in Exercise 2.6.5).

Proposition 2.4.31. *The Farey map F is ergodic with respect to the measure ν_F.*

Proof. To shorten the notation in what follows, let us denote the Borel σ-algebra on $[0, 1]$ by $\mathcal{B}$ and the Borel σ-algebra on $[1/2, 1] =: A$ by $\mathcal{B}_A$.

It was shown in Proposition 2.3.19 that the map F preserves the σ-finite Borel measure ν_F on the unit interval which is defined by the density h_F, given by $h_F(x) = 1/x$. It therefore follows from Proposition 2.4.29 that F_A preserves the measure $\nu_{F|_A}$.

We will now show that the induced system $(A, \mathcal{B}_A, \nu_{F|_A}, F|_A)$ of the Farey map and the Gauss system $([0, 1], \mathcal{B}, m_G, G)$ are measure-theoretically isomorphic. Recall that this means that there exist sets $X \subseteq [1/2, 1]$ and $Y \subseteq [0, 1]$ such that $\nu_F(X) = m_G(Y) = 1$ and a measure-preserving function $\psi : X \to Y$ such that $\psi \circ F_A = G \circ \psi$. Here, remember that the function ψ being measure-preserving means that $\nu_{F|_A} \circ \psi^{-1}(B) = m_G(B)$, for all $B \in \mathcal{B}$. Indeed, it suffices to let $X = [1/2, 1)$, $Y = (0, 1]$ and the function ψ be equal to the right-hand branch of the Farey map itself, that is, let $\varphi(x) := F|_A(x)$ for all $x \in [1/2, 1)$. Then, if $x = [1, x_2, x_3, \dots] \in A$, we have that

$$F|_A \circ F_A(x) = F|_A([1, x_3, x_4, \dots]) = [x_3, x_4, \dots] = G([x_2, x_3, \dots]) = G \circ F|_A(x).$$

Furthermore,

$$\nu_{F|_A}((F|_A)^{-1}([a, b])) = \nu_{F|_A}\left(\left[\frac{1}{1+b}, \frac{1}{1+a}\right]\right) = \frac{1}{\log 2} \int\limits_{1/(1+b)}^{1/(1+a)} \frac{1}{x}\, dx$$

$$= \frac{1}{\log 2} \cdot \log\left(\frac{b+1}{a+1}\right) = m_G([a, b]).$$

It is clear that two measure-theoretically isomorphic systems are either both ergodic or both not ergodic. Since we already know that G is ergodic, it follows from the argument above that F_A is also ergodic with respect to $\nu_{F|_A}$. Therefore, we can use Proposition 2.4.28 to conclude that the map F is ergodic with respect to ν_F. □

Remark 2.4.32. An argument similar to the one we have just given for F can be used to show that each α-Farey map F_α is ergodic with respect to the invariant measure discovered in Proposition 2.3.20. We leave the details to Exercise 2.6.9.

2.4.5 Uniqueness of the invariant measures for F and F_α

We will now turn to the question of uniqueness of Lebesgue-absolutely continuous invariant measures. It will turn out that for rather general systems there is, up to multiplication by a constant, only one λ-absolutely continuous invariant measure. The proof will employ the induced maps introduced in the previous section. We will also need the following proposition.

Proposition 2.4.33. Let $(X, \mathcal{B}, \mu, T)$ be a conservative and ergodic measure-preserving system, and for all $A \in \mathcal{B}$ with $0 < \mu(A) < \infty$, let φ denote the return time function on the set A. Then,

(a) $\mu(B) = \sum_{n=0}^{\infty} \mu(A \cap \{\varphi > n\} \cap T^{-n}(B))$, for all $B \in \mathcal{B}$.

(b) $\mu(X) = \int_A \varphi \, d\mu$.

Proof. By Lemma 2.4.3, every set A satisfying $0 < \mu(A) < \infty$ is a sweep-out set for T. Therefore, the function φ is well defined. Observe that for all $n \geq 0$,

$$T^{-1}(A^c \cap \{\varphi > n\}) = (A \cap \{\varphi > n+1\}) \cup (A^c \cap \{\varphi > n+1\}). \qquad (2.10)$$

(The proof of this fact is left to Exercise 2.6.8.) Now suppose that for all $B \in \mathcal{B}$ and some fixed $n \in \mathbb{N}$, we have

$$\mu(B) = \sum_{k=0}^{n} \mu(A \cap \{\varphi > k\} \cap T^{-k}(B)) + \mu(A^c \cap \{\varphi > n\} \cap T^{-n}(B)). \qquad (2.11)$$

Then, using (2.10) and the T-invariance of μ, we obtain that

$$\mu(B) = \sum_{k=0}^{n} \mu(A \cap \{\varphi > k\} \cap T^{-k}(B)) + \mu(T^{-1}(A^c \cap \{\varphi > n\}) \cap T^{-(n+1)}(B))$$

$$= \sum_{k=0}^{n} \mu(A \cap \{\varphi > k\} \cap T^{-k}(B))$$

$$\qquad + \mu(((A \cap \{\varphi > n+1\}) \cup (A^c \cap \{\varphi > n+1\})) \cap T^{-(n+1)}(B))$$

$$= \sum_{k=0}^{n+1} \mu(A \cap \{\varphi > k\} \cap T^{-k}(B)) + \mu(A^c \cap \{\varphi > n+1\} \cap T^{-(n+1)}(B)).$$

Consequently, since the formula is obviously true for $n = 0$, we have shown by induction that (2.11) holds for all $n \geq 0$. In order to finish the proof of part (a), we must show that $\lim_{n \to \infty} \mu(A^c \cap \{\varphi > n\} \cap T^{-n}(B)) = 0$. In order to do this, we split the remainder of the proof up into three cases.

(i) Let $B = T^{-1}(A)$. Then,

$$\mu(A) = \mu(T^{-1}(A))$$

$$= \sum_{k=0}^{n} \mu(A \cap \{\varphi > k\} \cap T^{-(k+1)}(A))$$

$$\qquad + \mu(A^c \cap \{\varphi > n\} \cap T^{-(n+1)}(A))$$

$$= \sum_{k=0}^{n} \mu(A \cap \{\varphi = k+1\}) + \mu(A^c \cap \{\varphi = n+1\}),$$

and since we can write $\mu(A) = \sum_{k=0}^{\infty} \mu(A \cap \{\varphi = k\}) < \infty$, we have that $\lim_{n \to \infty} \mu(A^c \cap \{\varphi > n\} \cap T^{-n}(B)) = 0$, as required.

Incidentally, notice that the above calculation also shows that, for all $n \geq 0$,

$$\mu(A^c \cap \{\varphi = n+1\}) = \mu(A) - \sum_{k=0}^{n} \mu(A \cap \{\varphi = k+1\})$$

$$= \mu\left(A \setminus \bigcup_{k=0}^{n}(A \cap \{\varphi = k+1\})\right)$$

$$= \mu(A \cap \{\varphi > n+1\}).$$

This will be useful in case (iii), below.

(ii) Suppose now that $B \subseteq A$. Then, $A^c \cap \{\varphi = n\} \cap T^{-n}(A) = \varnothing$, so in this case the proof is finished.

(iii) Finally, let $B \subseteq A^c \cap \{\varphi = N\}$, for some fixed $N \in \mathbb{N}$. In this case, we have that $A^c \cap \{\varphi > n\} \cap T^{-n}(B) \subseteq A^c \cap \{\varphi = n+N\}$. Thus, we have that

$$\lim_{n \to \infty} \mu(A^c \cap \{\varphi > n\} \cap T^{-n}(B)) = \lim_{n \to \infty} \mu(A^c \cap \{\varphi = n+N\})$$

$$= \lim_{n \to \infty} \mu(A \cap \{\varphi > N+n\}) = 0,$$

where the last two equalities are due to the observation made at the end of case (i) and the fact that the set A is assumed to be of finite measure. This finishes the proof of case (iii) and so completes the proof of part (a).

For part (b), if we substitute X for B into part (a), we obtain that

$$\mu(X) = \sum_{n=0}^{\infty} \mu(A \cap \{\varphi > n\} \cap T^{-n}(X)) = \sum_{n=0}^{\infty} \mu(A \cap \{\varphi > n\}) = \int_A \varphi \, d\mu.$$

☐

Remark 2.4.34. The result in Proposition 2.4.33 (b) is known as *Kac's formula*.

Theorem 2.4.35. *Let $(X, \mathcal{B}, \mu, T)$ be a conservative, ergodic, non-singular (but not necessarily measure-preserving) system. Then, up to multiplication by a constant, there is at most one μ-absolutely continuous σ-finite T-invariant measure.*

Proof. Let m_1 and m_2 be two non-zero, T-invariant σ-finite measures that are both absolutely continuous with respect to μ. Then, let $B \in \mathcal{B}$ with $\mu(B) > 0$. Since T is conservative and ergodic, Lemma 2.4.3 implies that the set B is a sweep-out set for T with respect to the measure μ. That is, we have

$$\bigcup_{n=0}^{\infty} T^{-n}(B) = X \mod \mu.$$

Therefore, since $\mu(X \setminus \bigcup_{n=0}^{\infty} T^{-n}(B)) = 0$, we also have that

$$m_1\left(X \setminus \bigcup_{n=0}^{\infty} T^{-n}(B)\right) = 0 \text{ and } m_2\left(X \setminus \bigcup_{n=0}^{\infty} T^{-n}(B)\right) = 0.$$

In other words, the set B is also a sweep-out set for T with respect to m_1 and m_2. In particular, $m_1(B), m_2(B) > 0$, so the measures m_1 and m_2 are in fact in the same measure class as μ.

Now choose $A \in B$ such that $0 < m_1(A) < \infty$ and $0 < m_2(A) < \infty$. We may assume, without loss of generality, that $m_1(A) = m_2(A) = 1$. Then, the measures $m_1|_A$ and $m_2|_A$ are equivalent ergodic T-invariant probability measures for the dynamical system given by $T_A : (A, B_A) \rightarrow (A, B_A)$. Thus, according to Proposition 2.4.8, we have that $m_1 = m_2$ on B_A. The formula in Proposition 2.4.33 (a) then yields that $m_1 = m_2$ on all of B. □

Corollary 2.4.36. *Up to multiplication by a constant, the invariant measures ν_F for the Farey system $([0, 1], B, F)$ and ν_α for the α-Farey system $([0, 1], B, F_\alpha)$ are unique.*

Proof. First, both F and F_α are non-singular with respect to λ, since ν_F, ν_α and λ are in the same measure class. Then, as F and F_α are both conservative and ergodic (see Proposition 2.4.31 and Exercise 2.6.9), an application of Theorem 2.4.35 gives that both ν_F and ν_α are unique. □

2.4.6 Proof of Hopf's Ratio Ergodic Theorem

Now, we will turn our attention to a proof of Hopf's Ratio Ergodic Theorem. The proof we will shortly present is due originally to Zweimüller [Zwe04]. It exploits the idea of inducing in a way that will allow us to apply the finite measure version of Birkhoff's Pointwise Ergodic Theorem.

Before we begin the proof, let us first fix some notation. Throughout, the system (X, B, μ, T) is assumed to be conservative, ergodic and measure-preserving. For $f \in L_1(\mu)$, we denote ergodic sums for the system T by

$$S_n(f) := \sum_{j=0}^{n-1} f \circ T^j.$$

We let A be a sweep-out set for T and consider the induced transformation $T_A : A \rightarrow A$ on A. For a measurable function $h : A \rightarrow \mathbb{R}$, we denote the ergodic sums for the induced system by

$$S_n^A(h) := \sum_{j=0}^{n-1} h \circ T_A^j.$$

A particularly important example is given by

$$\varphi_n := S_n^A(\varphi) = \sum_{j=0}^{n-1} \varphi \circ T_A^j, \tag{2.12}$$

where $\varphi : A \to \mathbb{N}$ is the return time function on A. Note that for a specific $x \in A$, the j-th summand $\varphi \circ T_A^j(x)$ inside this sum is equal to the length of the j-th excursion of the orbit $(T^n(x))_{n \geq 0}$ to the set A. To have a more concrete idea of what this means, it helps to think in terms of continued fractions. So, if $x = [1, x_1, x_2, x_3, \ldots] \in A := [1/2, 1]$ and if F_A denotes the Farey map induced on A, then $\varphi_1(x) := \varphi(x) = x_1$, $\varphi_2(x) := \varphi(x) + \varphi(F_A(x)) = x_1 + x_2$, and so on; in general,

$$\varphi_n(x) := \sum_{i=1}^{n} x_i.$$

Notice also that, trivially, we have

$$S_n^A(A) := S_n^A(\mathbb{1}_A) = n, \quad \text{for all } n \geq 1.$$

The idea of chopping up the orbits of points under T into pieces corresponding to each excursion to the set A is a useful one. We can also apply this idea to obtain the *induced version of a function* $f : X \to \mathbb{R}$, by adding up the values of the function observed during the first excursion and then represent these as a single function.

Definition 2.4.37. For $f : X \to \mathbb{R}$, let the function $f^A : A \to \mathbb{R}$ be defined by

$$f^A(x) := \sum_{j=0}^{\varphi(x)-1} f \circ T^j(x).$$

The function f^A is referred to as the *induced version of f on A.*

Lemma 2.4.38. *For an integrable function $f : X \to \mathbb{R}$, the following hold.*
(a)

$$S_{\varphi_n}(f) = S_n^A(f^A), \quad \text{for all } n \in \mathbb{N}.$$

(b)

$$\int_X f \, d\mu = \int_A f^A \, d\mu.$$

Proof. To prove part (a), we observe that for any $n \in \mathbb{N}$, the section of orbit $x, T(x), \ldots, T^{\varphi_n(x)-1}(x)$ that determines the sum $S_{\varphi_n}(f)$ consists of n complete excursions to A (that is, $T^{\varphi_n(x)}(x) \in A$). Therefore, we have that

$$S_{\varphi_n}(f) = S_{\varphi_1}(f) + S_{\varphi_2 - \varphi_1}(f \circ T_A) + \cdots + S_{\varphi_n - \varphi_{n-1}}(f \circ T_A^{n-1})$$
$$= S_{\varphi}(f) + S_{\varphi \circ T_A}(f \circ T_A) + \cdots + S_{\varphi \circ T_A^{n-1}}(f \circ T_A^{n-1})$$
$$= S_{\varphi}(f) + (S_{\varphi}(f)) \circ T_A + \cdots + (S_{\varphi}(f)) \circ T_A^{n-1} = S_n^A(f^A).$$

To prove part (b), let $f := \mathbb{1}_B$ for some $B \in \mathcal{B}$. Using Proposition 2.4.33 (a), we then have that

$$\int_X \mathbb{1}_B \, d\mu = \mu(B) = \sum_{n=0}^{\infty} \mu(A \cap \{\varphi > n\} \cap T^{-n}(B))$$

$$= \int_A \left(\sum_{n=0}^{\infty} \mathbb{1}_{A \cap \{\varphi > n\}} \cdot \mathbb{1}_B \circ T^n \right) d\mu$$

$$= \int_A \left(\sum_{n=0}^{\varphi-1} \mathbb{1}_B \circ T^n \right) d\mu = \int_A (\mathbb{1}_B)^A \, d\mu.$$

Hence, the assertion in part (b) holds for characteristic functions. A standard argument from measure theory then finishes the proof; we leave the details as an exercise. $\qquad\square$

Remark 2.4.39. The latter proposition also yields Kac's formula (see Remark 2.4.34) as a corollary, by simply choosing $f := \mathbb{1}_X$. In particular, note that this formula implies that the T-invariant measure μ is infinite if and only if the return-time function to any set A of positive finite measure is non-integrable.

We are now in a position to provide a proof of Hopf's Ratio Ergodic Theorem.

Proof of Theorem 2.4.24. Let A be a sweep-out set for T. First observe that it suffices to prove that for all $f \in L_1(\mu)$, we have that

$$\lim_{n \to \infty} \frac{S_n(f)}{S_n(\mathbb{1}_A)}(x) = \frac{\int_X f \, d\mu}{\mu(A)}, \quad \text{for } \mu\text{-a.e. } x \in A. \tag{2.13}$$

Indeed, the set of points where this limit exists and is equal to the right-hand side of the equality in (2.13) is T-invariant and of strictly positive μ-measure (since $\mu(A) > 0$). Therefore, the correct limit must be attained μ-a.e., by ergodicity. Then, if the same assertion is made for $g \in L_1(\mu)$, with the extra conditions that $g \geq 0$ and $\int_X g \, d\mu > 0$, the assertion of the theorem follows immediately.

Therefore, we are left only to give a proof of (2.13). For this, consider the induced map T_A. In light of Proposition 2.4.29, we have that T_A is an ergodic measure-preserving transformation on the finite measure space $(A, \mathcal{B}_A, \mu|_A)$. We can therefore apply Birkhoff's Pointwise Ergodic Theorem to T_A and the induced function f^A, which is integrable by Lemma 2.4.38, to deduce that

$$\lim_{n \to \infty} \frac{S_{\varphi_n}(f)}{S_{\varphi_n}(\mathbb{1}_A)} = \lim_{n \to \infty} \frac{S_n^A(f^A)}{n} = \int f^A \, d\mu|_A = \frac{\int_X f \, d\mu}{\mu(A)}, \quad \mu\text{-a.e. on } A. \tag{2.14}$$

This proves (2.13) for μ-a.e. $x \in A$ for the subsequence $(\varphi_n(x))_{n \geq 1}$. It remains to demonstrate convergence for the full sequence.

By the linearity of the integral, we may assume without loss of generality that $f \geq 0$. Then the sequence $(S_n(f))_{n \geq 1}$ is non-decreasing in n. Now, for a.e. $x \in A$ we find for every $k \in \mathbb{N}$ a positive integer n such that $\varphi_{n-1}(x) \leq k < \varphi_n(x)$. Therefore, observing that $S_k(\mathbb{1}_A)(x) = n - 1$ and using Lemma 2.4.38 (a), we have

$$\frac{S_{n-1}^A(f^A)(x)}{n-1} \leq \frac{S_k(f)(x)}{S_k(\mathbb{1}_A)(x)} \leq \frac{n}{n-1} \frac{S_n^A(f^A)(x)}{n}.$$

Observing that n tends to infinity as k tends to infinity, the proof is finished. $\qquad\square$

2.5 Exactness revisited

Recall from Definition 2.4.10 that a non-singular transformation T of a σ-finite measure space $(X, \mathcal{B}, \mu)$ is said to be exact if for each B in the tail σ-algebra $\bigcap_{n \in \mathbb{N}} T^{-n}(\mathcal{B})$ we have that either $\mu(B)$ or $\mu(X \setminus B)$ vanishes. In this section, our first aim is to prove the exactness of the Farey map. The second aim will be to give a useful equivalent formulation of exactness, known as *Lin's criterion*.

Let us first give a different characterisation of exactness, which we will employ to prove that the Farey map is exact. The origin of this characterisation is a paper by Miernowski and Nogueira [MN13] but it can also be found formulated more generally, in [Len14]. Similar, although not equivalent, ideas can be found already in a paper from the 1960s by Rokhlin [Rok64], where exactness was introduced for the first time. We will show that exactness of a transformation is implied by the following intersection property.

Definition 2.5.1. Let $(X, \mathcal{B}, \mu)$ be a σ-finite measure space and let $T : (X, \mathcal{B}, \mu) \to (X, \mathcal{B}, \mu)$ denote a bi-measurable map (that is, T is measurable and $T(B) \in \mathcal{B}$ for all $B \in \mathcal{B}$). Then T is said to satisfy the *intersection property* with respect to the measure μ provided that for every $A \in \mathcal{B}$ with positive measure, there exists some $k \geq 1$, depending on A, such that $\mu(T^k(A) \cap T^{k+1}(A)) > 0$.

Lemma 2.5.2. *Let the map* $T : (X, \mathcal{B}, \mu) \to (X, \mathcal{B}, \mu)$ *be bi-measurable and ergodic with respect to* μ. *If* T *satisfies the intersection property, then* T *is exact.*

Proof. Suppose that T is bi-measurable, ergodic and satisfies the intersection property, and let $A \in \bigcap_{m \in \mathbb{N}} T^{-m}(\mathcal{B})$. Suppose that $\mu(A) > 0$. In order to show that T is exact, we have to show that the complement of A has μ-measure equal to zero. Since A belongs to the tail σ-algebra, we have that $T^{-m}(T^m(A)) = A$, for all $m \geq 0$. We then have for all $m \geq 0$,

$$T^{m+1}(T^{-1}(A) \setminus A) = T^{m+1}(T^{-1}(T^{-m}(T^m(A))) \setminus T^{-(m+1)}(T^{m+1}(A)))$$
$$= T^{m+1}(T^{-(m+1)}(T^m(A)) \setminus T^{-(m+1)}(T^{m+1}(A)))$$
$$= T^m(A) \setminus T^{m+1}(A).$$

Using this, it follows for all $m \geq 1$, that

$$T^m(T^{-1}(A) \setminus A) \cap T^{m+1}(T^{-1}(A) \setminus A)$$
$$= (T^{m-1}(A) \setminus T^m(A)) \cap (T^m(A) \setminus T^{m+1}(A)) = \varnothing.$$

In particular, this shows that $\lambda\left(T^m\left(T^{-1}(A) \setminus A\right) \cap T^{m+1}\left(T^{-1}(A) \setminus A\right)\right) = 0$, for all $m \geq 1$. Hence, by the intersection property, we have that $\mu(T^{-1}(A) \setminus A) = 0$. Proceeding similarly for the set $A \setminus T^{-1}(A)$, we also obtain that $\mu(A \setminus T^{-1}(A)) = 0$. Hence, it follows that $T^{-1}(A) = A \mod \mu$. From there, the result is an immediate consequence of the ergodicity of T. $\qquad\square$

We will now begin to work towards the proof that F is exact. This will follow from a proposition given below, after we prove the following preparatory lemma. Let us remark that the main idea of the proof of exactness of F which we present here, including how to utilise the intersection property given in Definition 2.5.1, is inspired by the ideas of Lenci in [Len12], where one finds a similar proof valid for more general systems.

Lemma 2.5.3. *Consider the Farey system* $([0,1], \mathcal{B}, F)$, *and let A be given such that* $\lambda(A) > 0$. *Then*

$$\limsup_{n \to \infty} \lambda\left(F^n(A) \cap C(1)\right) = \lambda(C(1)).$$

Proof. We always have $\lambda\left(F^n(A) \cap C(1)\right) \leq \lambda(C(1))$. Hence we are left to show that there exists a strictly increasing sequence of positive integers $(n_k)_{k \geq 1}$ such that $\lim_{k \to \infty} \lambda\left(F^{n_k}(A) \cap C(1)\right) = \lambda(C(1))$. For this let $x = \langle x_1, x_2, x_3, \dots \rangle$ be a Lebesgue-density point[2] of A and recall that $(\widehat{C}(x_1, \dots, x_n))_{n \geq 1}$ denotes the shrinking family of Farey cylinder sets each containing x. Note that, since F is ergodic, (or by using Halmos's Recurrence Theorem), we can certainly choose x such that there exists a sequence $(n_k)_{k \geq 1}$ such that $x_{n_k+1} = 1$, for all $k \in \mathbb{N}$. To shorten the notation, let us define for all $k \in \mathbb{N}$,

$$D_k := \widehat{C}(x_1, \dots, x_{n_k}, x_{n_k+1}) = \widehat{C}(x_1, \dots, x_{n_k}, 1).$$

We have that F^{n_k} is bijective on D_k and we have that $F^{n_k}(D_k) = \widehat{C}(1) = C(1)$. Since x is a Lebesgue-density point of A, it follows that

$$\lim_{k \to \infty} \frac{\lambda(A \cap D_k)}{\lambda(D_k)} = 1 \text{ and } \lim_{k \to \infty} \frac{\lambda(D_k \setminus A)}{\lambda(D_k)} = 0. \tag{2.15}$$

---

2 A good reference for the Lebesgue density theorem and Lebesgue density points is Rudin [Rud87]; see in particular Theorem 7.2.

By partitioning $C(1)$, we have

$$1 = \frac{\lambda(C(1))}{\lambda(C(1))} = \frac{\lambda(C(1) \setminus F^{n_k}(A))}{\lambda(C(1))} + \frac{\lambda(F^{n_k}(A) \cap C(1))}{\lambda(C(1))}.$$

Here, the limit for k tending to infinity of the first summand in the above expression is equal to zero. Indeed, by first using the fact that $F^{n_k}(D_k) = C(1)$ and then Lemma 2.4.13 and (2.15), it follows that we have

$$\frac{\lambda(C(1) \setminus F^{n_k}(A))}{\lambda(C(1))} = \frac{\lambda(F^{n_k}(D_k \setminus A))}{\lambda(C(1))} \asymp \lambda(F^{n_k}(D_k \setminus A)) \asymp \frac{\lambda(D_k \setminus A)}{\lambda(D_k)} \to 0,$$

for k tending to infinity. This finishes the proof. $\qquad\square$

Proposition 2.5.4. *Let A be given such that $\lambda(A) > 0$. Then for the Farey map F we have that*

$$\limsup_{n \to \infty} \lambda(F^n(A) \cap F^{n+1}(A) \cap C(1)) = \lambda(C(1)).$$

Proof. Obviously, $\limsup_{n \to \infty} \lambda(F^n(A) \cap F^{n+1}(A) \cap C(1)) \leq \lambda(C(1))$. As in the proof of Lemma 2.5.3, let $x = \langle x_1, x_2, \dots \rangle$ be a Lebesgue-density point of A. In light of Remark 2.4.18, we have that there exists a sequence $(m_k)_{k \geq 1}$ such that $x_{m_k+1} = x_{m_k+2} = 1$, for all k. Therefore, for both of the sequences $(\widehat{C}(x_1, \dots, x_{m_k}, x_{m_k+1}))_{k \geq 1}$ and $(\widehat{C}(x_1, \dots, x_{m_k}, x_{m_k+1}, x_{m_k+2}))_{k \geq 1}$ we can proceed as in the proof of Lemma 2.5.3, which yields that

$$\lim_{k \to \infty} \lambda(F^{m_k}(A) \cap C(1)) = \lim_{k \to \infty} \lambda(F^{m_k+1}(A) \cap C(1)) = \lambda(C(1)).$$

Using this observation we obtain for the lower bound

$$\lambda(F^{m_k}(A) \cap F^{m_k+1}(A) \cap C(1))$$
$$= \lambda(F^{m_k}(A) \cap C(1)) + \lambda(F^{m_k+1}(A) \cap C(1)) - \lambda((F^{m_k}(A) \cup F^{m_k+1}(A)) \cap C(1))$$
$$\geq \lambda(F^{m_k}(A) \cap C(1)) + \lambda(F^{m_k+1}(A) \cap C(1)) - \lambda(C(1)) \to \lambda(C(1)),$$

for k tending to infinity. This proves the proposition. $\qquad\square$

Corollary 2.5.5. *For the Farey system $([0, 1], \mathcal{B}, F)$, let A be given such that $\lambda(A) > 0$. Then there exists $n \in \mathbb{N}$ such that*

$$\lambda(F^n(A) \cap F^{n+1}(A)) > 0.$$

In other words, we have that the Farey map F satisfies the intersection property with respect to the Lebesgue measure λ.

Theorem 2.5.6. *The Farey map F is exact with respect to the infinite invariant measure ν_F.*

Proof. Recalling that F is ergodic, this follows immediately on combining Corollary 2.5.5 with Lemma 2.5.2 and using the fact that λ and ν_F are in the same measure class. $\qquad\square$

Let us now move on to our second goal, the statement and proof of Lin's equivalent formulation of exactness [Lin71]. We will use this result in Chapter 5 to obtain information about certain sets defined in terms of their continued fraction expansion. In the proof here, we will make use of the dual space of $L_1(X, \mathcal{B}, \mu)$, which we already introduced in Remark 2.3.8, and the Banach–Alaoglu Theorem, which can be found, for instance, as Theorem V.4.2 in Dunford and Schwartz [DS88], if the reader has not encountered it before.

Theorem 2.5.7 (Lin's Criterion for Exactness). *Let $(X, \mathcal{B}, \mu, T)$ be a measure-preserving system. Then, T is exact if and only if for all $f \in L_1(X, \mathcal{B}, \mu)$ with $\int_X f \, d\mu = 0$ we have for the transfer operator $\widehat{T}$ that*

$$\lim_{n\to\infty} \left\| \widehat{T}^n(f) \right\|_1 = 0.$$

Proof. First suppose that T is exact and that $f \in L_1(X, \mathcal{B}, \mu)$ has zero expectation, that is, suppose that $\int_X f \, d\mu = 0$. Then, since $\|\widehat{T}\| = 1$ (see Exercise 2.6.11), the sequence $(\|\widehat{T}^n(f)\|_1)_{n\geq 1}$ is bounded. To show that its limit is zero, fix a subsequence $(n_k)_{k\geq 1}$ such that

$$\lim_{k\to\infty} \left\| \widehat{T}^{n_k}(f) \right\|_1 = \limsup_{n\to\infty} \left\| \widehat{T}^n(f) \right\|_1 < \infty.$$

If $(g_n)_{n\geq 1}$ is defined by

$$g_n := \operatorname{sign}\left(\widehat{T}^n(f) \right) \in L_\infty(X, \mathcal{B}, \mu),$$

then we have, for all $n \in \mathbb{N}$,

$$\left\| \widehat{T}^n(f) \right\|_1 = \int g_n \cdot \widehat{T}^n(f) \, d\mu = \int g_n \circ T^n \cdot f \, d\mu.$$

Since $\|g_n \circ T^n\|_\infty = 1$, it follows from the Riesz Representation Theorem that we can identify each $g_n \circ T^n$ with a bounded linear functional on $L_1(X, \mathcal{B}, \mu)$, that is, with an element in $L_1(X, \mathcal{B}, \mu)^* \simeq L_\infty(X, \mathcal{B}, \mu)$. By the Banach–Alaoglu Theorem we know that the closed unit ball in $L_1(X, \mathcal{B}, \mu)^*$ is compact in the weak-$*$ topology. Since $T^{-n-1}\mathcal{B} \subset T^{-n}\mathcal{B}$, it follows that $L_1\left(X, T^{-n-1}\mathcal{B}, \mu\right)^* \subset L_1\left(X, T^{-n}\mathcal{B}, \mu\right)^*$. Now for each $K \in \mathbb{N}$, we consider the non-empty and weak-$*$ compact sets G_K of accumulation points of the sequence $\left(g_{n_k} \circ T^{n_k}\right)_{k\geq K}$ in $L_1(X, \mathcal{B}, \mu)^*$, that is,

$$G_K := \left\{ g \in L_1\left(X, T^{-n_K}\mathcal{B}, \mu\right)^* : g \text{ is an accumulation point of } \left(g_{n_k} \circ T^{n_k}\right)_{k\geq K} \right\},$$

as subsets of the weak-$\star$ compact unit ball in $L_1(X, \mathcal{B}, \mu)^\star$. Since $G_K \subset G_{K+1}$ for all $K \in \mathbb{N}$, the intersection property of compact sets implies that $\bigcap_{K \in \mathbb{N}} G_K \neq \varnothing$. Fix $g \in \bigcap_{K \in \mathbb{N}} G_K$. By definition, $g \in L_1\left(X, T^{-n_K}\mathcal{B}, \mu\right)^\star \simeq L_\infty\left(X, T^{-n_K}\mathcal{B}, \mu\right)$ for all $K \in \mathbb{N}$, so we have that g is measurable with respect to the tail-σ-algebra $\bigcap_{n \in \mathbb{N}} T^{-n}\mathcal{B}$. Therefore, by the exactness of T, we have that g must be constant μ-a.e., that is, $g = c \in [0, \infty)$ μ-a.e., for some $c \in \mathbb{R}$. Since g is an accumulation point of the sequence $\left(g_{n_k} \circ T^{n_k}\right)_{k \geq 1}$, there exists a subsequence $\left(n_{k_\ell}\right)_{\ell \geq 1}$ such that $\lim_{\ell \to \infty} \int g_{n_{k_\ell}} \circ T^{n_{k_\ell}} \cdot f \, d\mu = \int g \cdot f \, d\mu$. This gives that

$$0 \leq \limsup_{n \to \infty} \left\|\widehat{T}^n(f)\right\|_1 = \lim_{k \to \infty} \left\|\widehat{T}^{n_k}(f)\right\|_1 = \lim_{k \to \infty} \int g_{n_k} \circ T^{n_k} \cdot f \, d\mu$$

$$= \lim_{\ell \to \infty} \int g_{n_{k_\ell}} \circ T^{n_{k_\ell}} \cdot f \, d\mu = \int g \cdot f \, d\mu = c \cdot \int f \, d\mu = 0.$$

In order to prove the converse, we assume that T is not exact and construct $f \in L_1(X, \mathcal{B}, \mu)$ with $\int f \, d\mu = 0$ and $\liminf_{n \to \infty} \left\|\widehat{T}^n(f)\right\|_1 > 0$. To that end, choose $A \in \bigcap_{n \in \mathbb{N}} T^{-n}\mathcal{B}$ such that $0 < \mu(A) < \mu(X)$, which is possible by the σ-finiteness of μ. For the same reason there exists a measurable set $B \subset X \setminus A$ such that $0 < \mu(B) < \infty$. For $f := \mathbb{1}_A - \mu(A)/\mu(B)\,\mathbb{1}_B$, we have that $f \in L_1(X, \mathcal{B}, \mu)$, $\int f \, d\mu = 0$ and $\int_A f \, d\mu = \mu(A) > 0$. Since $A \in \bigcap_{n \in \mathbb{N}} T^{-n}\mathcal{B}$, there exists a sequence $(A_n)_{n \geq 1}$ in $\mathcal{B}$ such that $A = T^{-n}A_n$, for each $n \in \mathbb{N}$. This yields that for all $n \in \mathbb{N}$, we have

$$\left\|\widehat{T}^n(f)\right\|_1 \geq \int_{A_n} \left|\widehat{T}^n f\right| d\mu \geq \int_{A_n} \widehat{T}^n f \, d\mu = \int \mathbb{1}_{A_n} \widehat{T}^n f \, d\mu$$

$$= \int \mathbb{1}_{A_n} \circ T^n f \, d\mu = \int_A f \, d\mu > 0.$$

This finishes the proof. □

We end this section by showing that if the underlying measure is finite, then exactness of a transformation implies the following mixing property. Discussion of mixing for infinite systems, whilst certainly interesting, is a much trickier business and we shall not go into it here. The interested reader is referred to the work of Lenci, see [Len14], [Len13], and references therein.

Definition 2.5.8. Let $(X, \mathcal{B}, \mu, T)$ be a measure-preserving system such that $\mu(X) = 1$. Then T is said to be *mixing* with respect to the measure μ provided that for every $A, B \in \mathcal{B}$, we have that

$$\lim_{n \to \infty} \mu(A \cap T^{-n}(B)) = \mu(A)\mu(B).$$

Corollary 2.5.9. *Let $(X, \mathcal{B}, \mu, T)$ be a measure-preserving system such that $\mu(X) = 1$. If T is exact, then T is mixing with respect to the measure μ.*

Proof. First note that since T preserves the probability measure μ, we have that $\mu \circ T^{-1} = \mu$, which implies $d(\mu \circ T^{-1})/d\mu = 1$ and hence, $\widehat{T}1 = 1$. Using this, it follows that for each $f \in L_1(X, \mathcal{B}, \mu)$ we have that $\widehat{T}(f - \int_X f\, d\mu) = \widehat{T}(f) - \int_X f\, d\mu$. Since $(f - \int_X f\, d\mu) \in L_1(X, \mathcal{B}, \mu)$ and $\int_X (f - \int_X f\, d\mu)d\mu = 0$, we can apply Lin's Criterion, which gives that $\lim_{n\to\infty} \|\widehat{T}^n(f - \int_X f\, d\mu)\|_1 = 0$. Using this and the fact that $1_A - \int_X 1_A\, d\mu \in L_1(X, \mathcal{B}, \mu)$, we obtain

$$\lim_{n\to\infty} \mu(A \cap T^{-n}(B)) = \lim_{n\to\infty} \int_X 1_A (1_B \circ T^n)\, d\mu = \lim_{n\to\infty} \int_X (\widehat{T}^n 1_A) 1_B\, d\mu$$

$$= \lim_{n\to\infty} \int_X (\widehat{T}^n(1_A - \mu(A)) 1_B\, d\mu + \mu(A)\mu(B)$$

$$= \mu(A)\mu(B).$$

This completes the proof. $\square$

2.6 Exercises

Exercise 2.6.1. Show that the Lebesgue measure is not invariant under the Gauss map.

Exercise 2.6.2. Show that the system $(\mathbb{R}, \mathcal{B}, \lambda, T)$ defined in Example 2.3.12 is really non-singular and that its Hopf decomposition is non-trivial with conservative part given by $\widehat{C}_T = [0, 1]$.

Exercise 2.6.3. As before, let $F_0 : x \mapsto x/(1+x)$ and $F_1 : x \mapsto 1/(1+x)$ denote the two inverse branches of the Farey map F. For $\omega = (\omega_1, \ldots, \omega_n) \in \{0, 1\}^n$, $n \in \mathbb{N}$, define

$$F_\omega := F_{\omega_1} \circ \ldots \circ F_{\omega_n},$$
$$\mathcal{E}_n := \{(\omega_1, \ldots, \omega_n) \in \{0, 1\}^n : \#\{i : \omega_i = 1\} \text{ is even}\}, \text{ and } \mathcal{O}_n := \{0, 1\}^n \setminus \mathcal{E}_n.$$

(i) Show that for all $n \in \mathbb{N}$ and $x \in (0, 1)$ we have that

$$x = \frac{\prod_{\omega \in \mathcal{E}_n} F_\omega(x)}{\prod_{\omega \in \mathcal{O}_n} F_\omega(x)}.$$

(ii) Use the identity in (i) to obtain an alternative proof of the fixed point equation $P_F(h) = h$ of the Ruelle operator P_F for the map F. (See Proposition 2.3.19).

Exercise 2.6.4. Give a proof of Proposition 2.3.21 (b), by verifying the eigenequation $P_{L_\alpha} h_{L_\alpha} = h_{L_\alpha}$ for the Ruelle operator P_{L_α} of an α-Lüroth map L_α.

Exercise 2.6.5. Let $F_A : A \to A$ denote the induced map of the Farey map F on the set $A := [1/2, 1]$. Prove that $F_A([x_1, x_2, x_3, x_4, \ldots]) = [x_1, x_3, x_4, \ldots]$, for all $[x_1, x_2, x_3, x_4, \ldots] \in A$.

Exercise 2.6.6. Let $(X, \mathcal{B}, \mu, T)$ be an ergodic system with a finite invariant measure μ, and suppose that $E \in \mathcal{B}$ is such that $\mu(E) > 0$. Let $(n_k)_{k \geq 0}$ be the sequence of occurrence times such that $T^{n_k}(x) \in E$ for all $k \geq 0$ (note that these are guaranteed to exist for μ-a.e. x by Halmos's Recurrence Theorem). Show that if we assume that $n_0 = 0$, so $x \in E$, then we have μ-a.e. that

$$\lim_{k \to \infty} \frac{n_k}{k} = \frac{1}{\mu(E)}.$$

Exercise 2.6.7. Prove that in an infinite ergodic system, the assumption of conservativity is necessary for the existence of sweep-out sets of finite measure. (See Lemma 2.4.3.)

Exercise 2.6.8. Prove that where $\varphi(x) := \inf\{n \geq 1 : T^n(x) \in A\}$ and A is a sweep-out set for T, we have

$$T^{-1}(A^c \cap \{\varphi > n\}) = (A \cap \{\varphi > n+1\}) \cup (A^c \cap \{\varphi > n+1\}), \quad \text{for all } n \geq 0.$$

Exercise 2.6.9. Taking inspiration from the proof of Proposition 2.4.31, prove that the map F_α is ergodic with respect to the measure ν_α defined in Proposition 2.3.20.

Exercise 2.6.10. Using the duality $(L_1(\mu))^* \simeq L_\infty(\mu)$ give a formal proof that the unitary operator $U_T : L_\infty(\mu) \to L_\infty(\mu), f \mapsto f \circ T$, is the dual operator of $\widehat{T}$.

Exercise 2.6.11. Prove that $\|\widehat{T}\| = 1$.

Exercise 2.6.12. Prove the statement in (2.7).

Exercise 2.6.13. Let μ and ν be two σ-finite measures on $(X, \mathcal{B})$ with $\mu \sim \nu$. Show that the Radon–Nikodým density $d\mu/d\nu$ is almost everywhere positive and that $d\nu/d\mu = (d\mu/d\nu)^{-1}$

Exercise 2.6.14. Show that if v/w is a reduced fraction in $(0, 1)$, then for all $p/q \in F^{-n}(v/w)$ we have that

$$\left| (F^n)' \left(\frac{p}{q} \right) \right| = \frac{q^2}{w^2}.$$

Exercise 2.6.15. Show that the statement in Proposition 2.5.4 still holds if we replace $\widehat{C}(1)$ by any arbitrary Farey-cylinder whose final symbol is equal to 1. (Of course, the sequence $(m_k)_{k \geq 1}$ might be a different one).

Exercise 2.6.16. Fill in the gap in the proof of Lin's Criterion: Prove that if g is measurable with respect to the tail σ-algebra of T and T is exact, then g is constant μ-a.e.

3 Renewal theory and α-sum-level sets

In this chapter we will mainly investigate certain subsets of the unit interval which are defined in terms of the α-Lüroth expansion. However, in order to motivate this exploration, in the first section we will describe the analogous problem for the continued fraction expansion. In the second section, we first define the sets we are interested in and then show how classical results in the field of renewal theory can be used to obtain detailed information about the sets in question.

3.1 Sum-level sets

One of the goals of Chapter 5 will be to give a detailed measure-theoretical analysis of the following sets C_n, for $n \in \mathbb{N}$, which we will call the *sum-level sets* for the continued fraction expansion:

$$C_n := \left\{ [x_1, x_2, x_3, \ldots] \in [0,1] : \sum_{i=1}^{k} x_i = n \text{ for some } k \in \mathbb{N} \right\}$$

The first few of these sets are shown in Fig. 3.1, below. Directly from the definition, we have that $C_1 = C(1) = [1/2, 1]$. Likewise, it follows that for the next few sum-level sets we have

$$C_2 = C(2) \cup C(1,1), \quad C_3 = C(3) \cup C(2,1) \cup C(1,1,1) \cup C(1,2),$$

and so on.

To begin the inspection of the sequence $(C_n)_{n \geq 1}$ of these sets, let us consider the *lim-inf set*, which is defined by

$$\liminf_{n \to \infty} C_n := \bigcup_{n \geq 1} \bigcap_{m \geq n} C_m = \{x \in [0,1] : x \in C_n \text{ for all sufficiently large } n\}.$$

In order for an irrational number x to lie in all of the sets $C_N, C_{N+1}, C_{N+2}, \ldots$, for some $N \in \mathbb{N}$, we must have that $x = [x_1, \ldots, x_k, 1, 1, 1, \ldots]$, where $\sum_{i=1}^{k} x_i = N$. In other words, the lim-inf set of the sequence $(C_n)_{n \geq 1}$ is equal to the set of all noble numbers (see Definition 1.1.13 (d)), that is, irrational numbers whose continued fraction digits are from some point on always equal to 1. As we have already observed, this set is

0   1

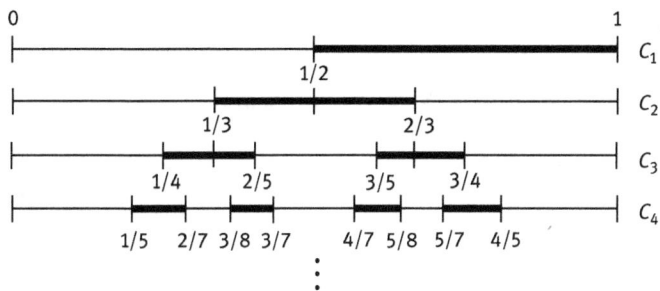

Fig. 3.1. The first four sum-level sets.

countable. On the other hand, one immediately verifies that the lim-sup set[1]

$$\limsup_{n\to\infty} C_n := \bigcap_{n\geq 1} \bigcup_{m\geq m} C_m = \{x \in [0,1] : x \in C_n \text{ for infinitely many } n\}$$

is equal to the set of all irrational numbers in $[0, 1]$. Hence, at first sight, the sequence of sum-level sets appears to be far away from being a canonical dynamical entity. (However, in Chapter 5 we will show that this is actually not the case.)

For the Lebesgue measure of the first four members of the sequence of the sum-level sets (cf. Fig. 3.1) one immediately computes that

$$\lambda(C_1) = 1/2, \quad \lambda(C_2) = 1/3, \quad \lambda(C_3) = 3/10 \text{ and } \lambda(C_4) = 39/140.$$

From this one might already start to suspect that the sequence $(\lambda(C_n))_{n\geq 1}$ is decreasing for n tending to infinity. In fact, it was conjectured by Fiala and Kleban [FK10] that $\lambda(C_n)$ tends to zero, as n tends to infinity. In Section 5.1, we will settle this conjecture affirmatively, as well as prove some much stronger results. Before this, though, we will consider the parallel, easier to analyse, situation for the α-Lüroth systems.

3.2 Sum-level sets for the α-Lüroth expansion

In this section, we will study the sequence of the Lebesgue measures of the α-sum-level sets for an arbitrary α-Lüroth map L_α, for a given partition α. Analogous to the sum-level sets for the continued fraction expansion, defined in the previous section,

---

[1] Recall that we have already encountered lim-sup sets in Chapter 1, specifically in Lemma 1.2.18 (the Borel–Cantelli Lemma).

these sets are given, for each $n \in \mathbb{N}$, by

$$C_n^{(\alpha)} := \left\{ x \in C_\alpha(\ell_1, \ell_2, \ldots, \ell_k) : \sum_{i=1}^{k} \ell_i = n, \text{ for some } k \in \mathbb{N} \right\}.$$

Also, for later convenience, we define $C_0^{(\alpha)} := [0, 1]$. Our toolkit for the investigation into the sequence $(\lambda(C_n^{(\alpha)}))_{n \geq 1}$ will consist of classical results from renewal theory.

3.2.1 Classical renewal results

Our aim here is to state and give some ideas of the proofs of some strong renewal theorems due to Garsia/Lamperti [GL63] and Erickson [Eri70]. Before doing so, we first state and prove the original discrete renewal theorem due to Erdős, Pollard and Feller [EFP49]. We begin by defining a renewal pair.

Definition 3.2.1. Let $(v_n)_{n \geq 1}$ be an infinite probability vector, that is, a sequence of non-negative real numbers for which $\sum_{k=1}^{\infty} v_n = 1$. Assume that associated to this vector there exists a sequence $(w_n)_{n \geq 0}$, with $w_0 := 1$, which satisfies the *renewal equation*:

$$w_n = \sum_{m=1}^{n} v_m w_{n-m}, \quad \text{for all } n \in \mathbb{N}.$$

A pair $((v_n)_{n \geq 1}, (w_n)_{n \geq 0})$ of sequences with these properties is referred to as a *renewal pair*.

Let us give a brief sketch of the original probabilistic motivation for this definition. For further details and many examples, we refer the reader to Chapter XIII of Feller [Fel68a].

Consider a sequence of independent identically distributed random variables $(T_n)_{n \in \mathbb{N}}$ with values in $\mathbb{N}$. (Just think of T_n as the random discrete time between the occurrence of a 'recurrent event', like the successive renewal of a burned-out lightbulb.) For the distribution, we write $v_k := \mathbb{P}(T_1 = k)$ for each $k \in \mathbb{N}$. Now the probability of the occurrence of the event at time $k \in \mathbb{N}$ is given by

$$w_k := \mathbb{P}\left(\left\{ \sum_{i=1}^{\ell} T_i = k \text{ for some } \ell \in \mathbb{N}_0 \right\} \right).$$

Since the empty sum is by definition equal to 0 we have $w_0 = 1$. Using the fact that the sequence of random variables (T_n) are independent and identically distributed we find

$$
w_n = \mathbb{P}\left(\left\{\sum_{i=1}^{\ell} T_i = n \text{ for some } \ell \in \mathbb{N}_0\right\}\right)
$$

$$
= \sum_{m=1}^{n} \mathbb{P}\left(\left\{T_1 = m \text{ and } \sum_{i=1}^{\ell} T_i = n \text{ for some } \ell \in \mathbb{N}\right\}\right)
$$

$$
= \mathbb{P}(\{T_1 = n\}) + \sum_{m=1}^{n-1} \mathbb{P}\left(\left\{T_1 = m \text{ and } \sum_{i=2}^{\ell} T_i = n - m \text{ for some } \ell \in \mathbb{N}\right\}\right)
$$

$$
= v_n + \sum_{m=1}^{n-1} v_m \mathbb{P}\left(\left\{\sum_{i=1}^{\ell} T_{i+1} = n - m \text{ for some } \ell \in \mathbb{N}_0\right\}\right)
$$

$$
= \sum_{m=1}^{n} v_m w_{n-m}.
$$

This shows that the renewal equation $w_n = \sum_{m=1}^{n} v_m w_{n-m}$ for $n \in \mathbb{N}$ has its natural place in probability theory. In the following we will see how to determine the asymptotic behaviour of (w_n) in terms of (v_n) just by analysing the renewal equation.

We are now almost in a position to state and prove the classical discrete renewal theorem. The proof we give here is essentially (with a few extra details inserted) the original proof given in [EFP49]. Before we start, we make the following definitions for a given renewal pair $((v_n), (w_n))$:

$$
d_v := \gcd\{n \geq 1 : v_n > 0\} \quad \text{and} \quad d_w := \gcd\{n \geq 1 : w_n > 0\}.
$$

Then, for all n with $w_n = 0$ we also have that $v_n = 0$, since using the renewal equation gives that $w_n = 0 = \sum_{m=1}^{n} v_m w_{n-m}$ and so each term in this sum must be equal to zero. In particular, $v_n w_0 = 0$, but since $w_0 = 1$, it follows that $v_n = 0$. This implies that d_w is a factor of d_v. It is also possible to show, using a fairly straightforward but somewhat ungainly induction argument, that d_v is a factor of d_w. Thus, these two quantities are always equal. We will also need the following elementary technical observation.

Lemma 3.2.2. *If d is the greatest common divisor of the sequence of natural numbers $(n_k)_{k \geq 1}$, then there exist numbers K and M with the property that for each $m \in \mathbb{N}$ such that $m \geq M$ there exist $c_1, \ldots, c_K \in \mathbb{N}$ such that*

$$
m \cdot d = \sum_{k=1}^{K} c_k n_k.
$$

Proof. We can assume that $d = 1$ (otherwise just divide each of the n_k by d), and also that d is the greatest common divisor of the first K of the given numbers, that is, g.c.d.$(n_1, n_2, \ldots, n_K) = 1$. It is well known that there then exist integers $b_1, b_2, \ldots, b_K$

with the property that[2]

$$b_1 n_1 + \ldots + b_K n_K = 1.$$

Letting $b := \max\{|b_1|, |b_2|, \ldots, |b_K|\}$ and $M := bn_1(n_1 + \cdots + n_K)$, we have that each $m \geq M$ can be written in the form

$$m = bn_1(n_1 + \cdots + n_K) + in_1 + r(b_1 n_1 + \cdots + b_K n_K),$$

where $i \geq 0$ and $0 \leq r < n_1$ come from the division algorithm applied to $m - M$. Therein lie the factors c_k and (since $bn_1 > b_k r$), they are clearly positive integers. $\qquad\square$

Finally, before stating the theorem, we also need the following elementary lemma. We include the proof for completeness.

Lemma 3.2.3. *Let $(b_n)_{n\geq 1}$ and $(b'_n)_{n\geq 1}$ be two sequences of real numbers with the property that $\lim_{n\to\infty}(b_n + b'_n)$ exists. Then, provided that we do not have $\liminf_{n\to\infty} b_n = -\infty$ and $\limsup_{n\to\infty} b'_n = \infty$, or vice versa, it follows that*

$$\lim_{n\to\infty}(b_n + b'_n) = \liminf_{n\to\infty} b_n + \limsup_{n\to\infty} b'_n.$$

Proof. On the one hand we have

$$\lim_{n\to\infty}(b_n + b'_n) - \limsup_{i\to\infty}(b'_n) = \liminf_{n\to\infty}(b_n + b'_n) + \liminf_{i\to\infty}(-b'_n) \leq \liminf_{n\to\infty}(b_n).$$

On the other hand,

$$\lim_{n\to\infty}(b_n + b'_n) - \liminf_{i\to\infty}(b_n) = \limsup_{n\to\infty}(b_n + b'_n) + \limsup_{i\to\infty}(-b_n) \geq \limsup_{n\to\infty}(b'_n).$$

Combining these two inequalities, the lemma is proved. $\qquad\square$

Theorem 3.2.4 (Discrete Renewal Theorem). *Let $((v_n)_{n\geq 1}, (w_n)_{n\geq 0})$ be a renewal pair and suppose that $d_v = 1$. Then*

$$\lim_{n\to\infty} w_n = \frac{1}{\sum_{m=1}^{\infty} m \cdot v_m},$$

where the limit is understood to be equal to zero if the series in the denominator diverges.

Proof. For ease of notation, throughout we denote $s := \sum_{m=1}^{\infty} m \cdot v_m$. First, we show by induction that $0 \leq w_n \leq 1$, for each $n \in \mathbb{N}_0$. To start, notice that $w_1 = v_1 \cdot w_0 = v_1 \leq 1$.

---

2 This can be seen, for instance, by considering moduli of integers. The interested reader is referred to Section 2.9 of The Theory of Numbers, by Hardy and Wright [HW08].

Now suppose that $0 \le w_k \le 1$ for all $0 \le k \le n - 1$. Then

$$w_n = v_1 \cdot w_{n-1} + v_2 \cdot w_{n-2} + \cdots + v_n \cdot w_0 \le \sum_{k=1}^{n} v_k \le 1.$$

Let $w := \limsup_{n \to \infty} w_n$ and pick a subsequence $(w_{n_k})_{k \in \mathbb{N}}$ with the property that $\lim_{k \to \infty} w_{n_k} = w$. Then, for all $m \ge 1$, we have, via Lemma 3.2.3, that

$$w = \lim_{k \to \infty} w_{n_k} = \lim_{k \to \infty} \left(v_m \cdot w_{n_k - m} + \sum_{\substack{1 \le s \le n_k \\ s \neq m}} v_s \cdot w_{n_k - s} \right)$$

$$= \liminf_{k \to \infty} v_m \cdot w_{n_k - m} + \limsup_{k \to \infty} \sum_{\substack{1 \le s \le n_k \\ s \neq m}} v_s \cdot w_{n_k - s}$$

$$\le v_m \liminf_{k \to \infty} w_{n_k - m} + \sum_{s \neq m} v_s \limsup_{k \to \infty} w_{n_k - s}$$

$$\le v_m \liminf_{k \to \infty} w_{n_k - m} + (1 - v_m) w.$$

From this it follows immediately that $v_m w \le v_m \liminf_{k \to \infty} w_{n_k - m}$ and therefore, provided that $v_m > 0$, we obtain

$$\limsup_{n \to \infty} w_n =: w \le \liminf_{k \to \infty} w_{n_k - m}.$$

Thus,

$$\lim_{k \to \infty} w_{n_k - m} = w. \tag{3.1}$$

Applying this argument many times over, one obtains that equation (3.1) holds for all m such that there exist positive integers $m_1, \ldots m_j$ with each $v_{m_i} > 0$ so that $m = m_1 + \ldots + m_j$. Given that $d_v = 1$, from Lemma 3.2.1 it follows that every large enough m has this form (where we can do without the factors c_i, as there is no reason that the integers m_i have to be distinct). In other words, there exists some $M \in \mathbb{N}$ such that (3.1) holds for every $m \ge M$.

Now, for each $n \in \mathbb{N}$, set

$$r_n := \sum_{m=n+1}^{\infty} v_m.$$

Then $r_0 = 1$ and

$$\sum_{m=1}^{\infty} m \cdot v_m = \sum_{m=1}^{\infty} v_m + \sum_{m=2}^{\infty} v_m + \sum_{m=3}^{\infty} v_m + \ldots = \sum_{n=0}^{\infty} r_n.$$

From the renewal equation and the fact that $r_m - r_{m-1} = -v_m$, we deduce that

$$r_0 \cdot w_n = w_n = -\sum_{m=1}^{n} (r_m - r_{m-1})w_{n-m}$$

and, by bringing the negative terms to the left-hand side, we can write this in the following way:

$$r_0 \cdot w_n + r_1 \cdot w_{n-1} + \cdots + r_n \cdot w_0 = r_0 \cdot w_{n-1} + \cdots + r_{n-1} \cdot w_0. \tag{3.2}$$

If we define the left-hand side of Equation (3.2) to be equal to s_n, then the right-hand side of Equation (3.2) is equal to s_{n-1}. Note that $s_0 = r_0 \cdot w_0 = 1$. Thus, in light of Equation (3.2), we have that $s_n = 1$, for all $n \in \mathbb{N}$. In particular,

$$\sum_{i=0}^{n_k-M} r_i \cdot w_{n_k-(M+i)} = 1. \tag{3.3}$$

We will now show that $w = 1/s$. First, suppose that s is finite. In that case, for all $\varepsilon > 0$ there exists $N \in \mathbb{N}$ with

$$r_0 + r_1 + \ldots r_N \geq s - \varepsilon.$$

If k is sufficiently large such that $n_k - M \geq N$, then by (3.3) we have

$$1 \geq \sum_{i=0}^{N} r_i \cdot w_{n_k-(M+i)}.$$

It then follows from (3.1) that

$$1 \geq w(r_1 + \ldots r_N) \geq w(s - \varepsilon).$$

Since ε was an arbitrary positive number, we obtain the inequality $w \leq 1/s$.

On the other hand, from (3.3) and from the pair of inequalities $w_n \leq 1$ and $(r_{N+1} + r_{N+2} + \ldots) \leq \varepsilon$, we deduce that

$$1 \leq \varepsilon + \sum_{i=0}^{N} r_i \cdot w_{n_k-(M+i)}.$$

Letting k tend to infinity, the above equation yields that

$$1 \leq \varepsilon + w \sum_{m=1}^{\infty} m \cdot v_m,$$

and so we also have the opposite inequality, namely, $w \geq 1/s$.

If we are instead in the situation that s is infinite, we have for all $C > 0$, that there exists an $N \in \mathbb{N}$ such that

$$r_0 + r_1 + \ldots r_N > C,$$

from which, in a similar way to the above, we obtain the inequality $1 \geq Cw$. Since C can be arbitrarily large, it follows that $w = 0$. Notice that if $\limsup_{n \to \infty} w_n = 0$, then we must have that $\lim_{n \to \infty} w_n = 0$, as these are all positive numbers. Therefore, in the case where s is infinite, the proof is finished.

In the case where s is finite, we also have to show that $\liminf_{n \to \infty} w_n = 1/s$. This proceeds analogously, starting by setting $w' := \liminf_{n \to \infty} w_n$ and then choosing a subsequence that achieves this lower limit. □

Remark 3.2.5. In the above proof, if it so happens that $v_m > 0$ for every $m \in \mathbb{N}$, we could dispense with the slight complication of having to use Lemma 3.2.2, since in this situation we have that Equation (3.1) holds for every $m \in \mathbb{N}$.

We will now state some stronger renewal results obtained by Garsia and Lamperti [GL63], and by Erickson [Eri70]. Their results are for the case where the limit in the statement of the discrete renewal theorem is equal to zero. They study the manner in which the sequence $(w_n)_{n \geq 0}$ tends to zero, under a certain additional hypothesis which we will now describe. Let the sequences $((v_n)_{n \geq 1}, (w_n)_{n \geq 0})$ be a given renewal pair and let the two associated sequences $(V_n)_{n \in \mathbb{N}}$ and $(W_n)_{n \in \mathbb{N}}$ be defined, for all $n \in \mathbb{N}$, by

$$V_n := \sum_{k=n}^{\infty} v_k \quad \text{and} \quad W_n := \sum_{k=1}^{n} w_k. \tag{3.4}$$

Then the principal assumption in these strong renewal results is that $V(n)$ satisfies

$$V_n = \psi(n) n^{-\theta},$$

for all $n \in \mathbb{N}$, for some $\theta \in [0, 1]$ and for some slowly varying function ψ. (Recall that slowly varying functions were defined in Section 1.4.4.) Before stating the theorem, let us also remind the reader that the notation "$f(n) \sim g(n)$" means that $\lim_{n \to \infty} f(n)/g(n) = 1$. Finally, the constants appearing on the right-hand side of the first two statements are given in terms of the gamma function (which was originally introduced by Euler). The gamma function is an extension of the factorial function to complex arguments, so we have $\Gamma(n) = (n-1)!$, and, considered as an extension to the open right-half plane, it has no zeros. For more details, we refer the interested reader to the book Complex Analysis by Gamelin [Gam01].

Strong renewal results by Garsia/Lamperti and Erickson [GL63, Lemma 2.3.1], [Eri70, Theorem 5]:

For $\theta \in [0, 1]$, we have that

$$W_n \sim (\Gamma(2-\theta)\Gamma(1+\theta))^{-1} \cdot n \cdot \left(\sum_{k=1}^{n} V_k\right)^{-1}.$$

Also, if $\theta \in (1/2, 1]$, then

$$w_n \sim (\Gamma(2-\theta)\Gamma(\theta))^{-1} \cdot \left(\sum_{k=1}^{n} V_k\right)^{-1}.$$

Finally, for $\theta \in (0, 1/2)$ we have that the limit in the latter formula does not have to exist in general. However, for $\theta \in (0, 1/2]$ it is shown in [GL63, Theorem 1.1] that one at least has

$$\liminf_{n\to\infty} n \cdot w_n \cdot V_n = \frac{1}{\Gamma(\theta)\Gamma(1-\theta)} = \frac{\sin \pi\theta}{\pi},$$

and that the limit exists if we restrict the indices to a set of integers whose complement is of zero density[3]

We will not rigorously prove these strong renewal results, as the proofs are decidedly non-trivial. However, we will provide a sketch of some of the main ideas. The proof of the first statement in the strong renewal results by Garsia/Lamperti and Erickson is reasonably straightforward, although it does use some fairly heavy analytic machinery. The deep result underlying this statement is *Karamata's Tauberian Theorem*, which we state below in the setting of power series (the proof can be found in [Fel68b]). Before stating this theorem, let us recall the following definitions:

- A measurable function $\psi : \mathbb{R}^+ \to \mathbb{R}^+$ is said to be *slowly varying* if

$$\lim_{x\to\infty} \frac{\psi(xy)}{\psi(x)} = 1, \text{ for all } y > 0.$$

- A function $f : \mathbb{R}^+ \to \mathbb{R}^+$ is called *regularly varying with exponent ρ* (with $\rho \in \mathbb{R}$) if for all $x \in \mathbb{R}^+$ we have

$$f(x) = x^\rho \cdot \psi(x),$$

where ψ is slowly varying.
- A sequence $(b_n)_{n\in\mathbb{N}}$ is called regularly varying with exponent ρ if for all $n \in \mathbb{N}$, we have that $b_n = f(n)$, with $f : \mathbb{R}^+ \to \mathbb{R}^+$ regularly varying with exponent ρ.

---

[3] If we set $A(n) := \{1, \ldots, n\} \cap A$, then the density of a set of integers A is given, where the limit exists, by $d(A) := \lim_{n\to\infty} \#A(n)/n$. For example, if $A := \{n^2 : n \in \mathbb{N}\}$, then since $\#A(n) \leq \sqrt{n}$ we have that $d(A) = 0$.

Theorem 3.2.6 (Karamata's Tauberian Theorem). *Let $b_n \geq 0$ for all $n \in \mathbb{N}_0$ and suppose that the series*

$$B(s) := \sum_{n=0}^{\infty} b_n s^n$$

converges for $0 \leq s < 1$. If ψ is slowly varying and $0 \leq \rho < \infty$, then the following two statements are equivalent:

(a) $B(s) \sim \dfrac{1}{(1-s)^\rho} \cdot \psi\left(\dfrac{1}{1-s}\right)$, *as $s \to 1^-$.*

(b) $\displaystyle\sum_{k=0}^{n-1} b_k \sim \dfrac{n^\rho \cdot \psi(n)}{\Gamma(1+\rho)}$, *as $n \to \infty$.*

Furthermore, if the sequence $(b_n)_{n\in\mathbb{N}}$ is monotonic and $0 < \rho < \infty$, then (a) is equivalent to

(c) $b_n \sim \dfrac{n^{\rho-1} \cdot \psi(n)}{\Gamma(\rho)}$, *as $n \to \infty$.*

Finally, if for a family of sequences $(b_n^x)_{x\in X}$ the asymptotic in (b) holds uniformly in $x \in X$ then so does the asymptotic in (a).

Proof. See [Fel68b], Theorem 5 in Section XIII.5 and, for the uniformity, a detailed inspection of the proof of the Extended Continuity Theorem (Section XIII.1 Theorem 2a) is needed (cf. Excercises 3.3.6 and 3.3.7). $\qquad\qquad\qquad\qquad\qquad\qquad$ □

For the following discussion, we will need to use the notion of a *generating function* (see also Chapter XI of [Fel68b]).

Definition 3.2.7. Let $(c_i)_{i\geq 0}$ be a sequence of real numbers and define

$$C(s) := \sum_{n=0}^{\infty} c_n s^n.$$

Then, if $C(s)$ is convergent in some interval $-s_0 < s < s_0$, we say that $C(s)$ is the *generating function* of the sequence $(c_i)_{i\geq 0}$. Note that if the sequence $(c_i)_{i\geq 0}$ is bounded, then $C(s)$ certainly converges in the interval $|s| < 1$.

Now, for a given renewal pair $((v_n)_{n\geq 1}, (w_n)_{n\geq 0})$, and with W_n and V_n defined as in (3.4), we wish to use Karamata's Tauberian Theorem to prove that

$$W_n \sim (\Gamma(2-\theta)\Gamma(1+\theta))^{-1} \cdot n \cdot \left(\sum_{k=1}^{n} V_k\right)^{-1}.$$

To begin, first notice that

$$\sum_{k=1}^{n} V_k \sim (1-\theta)^{-1} \cdot n^{1-\theta} \cdot \psi(n), \text{ as } n \to \infty.$$

Now, let us define two generating functions

$$V(t) := \sum_{n=0}^{\infty} V_n t^n \text{ and } W(t) := \sum_{n=0}^{\infty} w_n t^n.$$

Recalling both that $V_n = n^{-\theta} \psi(n)$, where $\theta \in [0, 1]$ and ψ is a slowly varying function, and that V_n is monotonically decreasing, we can apply Theorem 3.2.6 (c) $\Longrightarrow$ (a) to obtain

$$V(t) \sim \Gamma(1 - \theta)(1 - t)^{\theta - 1} \psi \left(\frac{1}{1 - t} \right).$$

Multiplying out and gathering coefficients yields that $V(t)W(t) = 1/(1 - t)$, or, in other words,

$$W(t) = \frac{1}{(1 - t)V(t)}.$$

Thus,

$$W(t) \sim \frac{1}{\Gamma(1 - \theta) \psi \left(\frac{1}{1 - t} \right)} (1 - t)^{-\theta}.$$

Finally, applying Theorem 3.2.6 (a) $\Longrightarrow$ (b), we have that

$$W_n \sim \frac{1}{\Gamma(1 - \theta)} \cdot n^\theta \cdot \frac{1}{\psi(n)} \cdot \frac{1}{\Gamma(1 + \theta)}$$

$$= \frac{1 - \theta}{\Gamma(1 + \theta)\Gamma(2 - \theta)} \cdot n^\theta \cdot \frac{1}{\psi(n)}$$

$$= \frac{1}{\Gamma(1 + \theta)\Gamma(2 - \theta)} \cdot n \cdot \left(\sum_{k=1}^{n} V_k \right)^{-1}.$$

The proof of the other parts of the strong renewal results quoted above also rely on Theorem 3.2.6, but also on some intricate estimates of integrals. We leave the details to the intrepid reader.

3.2.2 Renewal theory applied to the α-sum-level sets

We begin our discussion with the crucial observation that the sequence of the Lebesgue measures of these α-sum-level sets satisfies a renewal equation (as observed in [WX11], [Mun11] and [KMS12]). Here, the role of the probability vector is filled by the sequence of Lebesgue measures of the partition elements of α, that is, the sequence $(a_m)_{m \geq 1}$.

Lemma 3.2.8. *We have that* $\left(a_n, C_n^{(\alpha)}\right)$ *defines a renewal pair. That is, for each* $n \in \mathbb{N}$, *we have*

$$\lambda\left(C_n^{(\alpha)}\right) = \sum_{m=1}^{n} a_m \lambda\left(C_{n-m}^{(\alpha)}\right).$$

Proof. Since $\lambda(C_0^{(\alpha)}) = 1$ and $\lambda(C_1^{(\alpha)}) = a_1$, the assertion certainly holds for $n = 1$. For $n \geq 2$, the following calculation finishes the proof.

$$\lambda\left(C_n^{(\alpha)}\right) = \lambda\left(C_\alpha(n)\right) + \sum_{m=1}^{n-1} \sum_{\substack{C_\alpha(\ell_1,\ldots,\ell_k,m) \in C_n^{(\alpha)} \\ k \in \mathbb{N}}} \lambda(C_\alpha(\ell_1,\ldots,\ell_k,m))$$

$$= \lambda(C_\alpha(n)) + \sum_{m=1}^{n-1} a_m \sum_{\substack{C_\alpha(\ell_1,\ldots,\ell_k) \in C_{n-m}^{(\alpha)} \\ k \in \mathbb{N}}} \lambda(C_\alpha(\ell_1,\ldots,\ell_k))$$

$$= a_n \lambda\left(C_0^{(\alpha)}\right) + \sum_{m=1}^{n-1} a_m \lambda\left(C_{n-m}^{(\alpha)}\right) = \sum_{m=1}^{n} a_m \lambda\left(C_{n-m}^{(\alpha)}\right). \qquad \square$$

We are now in a position to prove our main results. The first of these is valid for arbitrary partitions, but for the second we must restrict ourselves to partitions that are either expansive of exponent $\theta \in [0, 1]$ or of finite type (recall that these were introduced in Definition 1.4.18). The proof of the first statement of the first main result again makes use of the notion of a generating function, which was defined above (see Definition 3.2.7).

Theorem 3.2.9. *For the α-sum-level sets of an arbitrary given partition $\alpha \in \mathcal{A}$ we have that $\sum_{n=1}^{\infty} \lambda(C_n^{(\alpha)})$ diverges, and that*

$$\lim_{n \to \infty} \lambda\left(C_n^{(\alpha)}\right) = \begin{cases} 0 & \text{if } F_\alpha \text{ is of infinite type;} \\ \left(\sum_{k=1}^{\infty} t_k\right)^{-1} & \text{if } F_\alpha \text{ is of finite type.} \end{cases}$$

Proof. The general form of the discrete renewal theorem given in Theorem 3.2.4 above can be applied directly to our specific situation. For this, fix some partition $\alpha \in \mathcal{A}$, and set $v_n := \lambda(A_n) = a_n$, for each $n \in \mathbb{N}$. Let us recall again that this is certainly a probability vector. Then, put $w_n := \lambda(C_n^{(\alpha)})$, for each $n \in \mathbb{N}_0$. In light of Lemma 3.2.8 and the observation that $w_0 = \lambda(C_0^{(\alpha)}) = 1$, we then have that these particular sequences $(v_n)_{n \geq 1}$ and $(w_n)_{n \geq 0}$ are indeed a renewal pair. Consequently, an application of the discrete renewal theorem immediately implies that

$$\lim_{n \to \infty} \lambda\left(C_n^{(\alpha)}\right) = \left(\sum_{k=1}^{\infty} k \cdot a_k\right)^{-1} = \left(\sum_{k=1}^{\infty} t_k\right)^{-1},$$

where this limit is equal to zero if $\sum_{k=1}^{\infty} t_k$ diverges. Note that by Lemma 2.3.20, the divergence of the latter series is equivalent to the statement that the partition α is of infinite type.

For the remaining assertion, let us consider the two generating functions V and W, which are given by

$$V(s) := \sum_{n=1}^{\infty} v_n s^n \text{ and } W(s) := \sum_{m=0}^{\infty} w_m s^m.$$

Using the Cauchy product formula for the two power series in tandem with the renewal equation provided in Lemma 3.2.8, we have that

$$W(s)V(s) = \sum_{n=1}^{\infty} s^n \sum_{m=1}^{n} v_m w_{n-m} = \sum_{n=1}^{\infty} w_n s^n = W(s) - 1 \tag{3.5}$$

Hence, $W(s) = 1/(1 - V(s))$. Since $a(1) = 1$, this yields that

$$\lim_{s \to 1^-} W(s) = +\infty,$$

which shows that the series $\sum_{n=0}^{\infty} \lambda(C_n^{(\alpha)})$ diverges. This finishes the proof. □

Theorem 3.2.10. *For a given partition α which is either expansive of exponent $\theta \in [0, 1]$ or of finite type, we have the following estimates for the asymptotic behaviour of the Lebesgue measure of the α-sum-level sets.*

(a) *With $K_\theta := (\Gamma(2 - \theta)\Gamma(1 + \theta))^{-1}$ for α expansive of exponent $\theta \in [0, 1]$ and with $K_\theta := 1$ for α of finite type, we have that*

$$\sum_{k=1}^{n} \lambda\left(C_k^{(\alpha)}\right) \sim K_\theta \cdot n \cdot \left(\sum_{k=1}^{n} t_k\right)^{-1}.$$

(b) *With $k_\theta := (\Gamma(2 - \theta)\Gamma(\theta))^{-1}$ for α expansive of exponent $\theta \in (1/2, 1]$ and with $k_\theta := 1$ for α of finite type, we have that*

$$\lambda\left(C_n^{(\alpha)}\right) \sim k_\theta \cdot \left(\sum_{k=1}^{n} t_k\right)^{-1}.$$

(c) *For an expansive partition α of exponent $\theta \in (0, 1)$, we have that*

$$\liminf_{n \to \infty} \left(n \cdot t_n \cdot \lambda\left(C_n^{(\alpha)}\right)\right) = \frac{\sin \pi \theta}{\pi}.$$

Moreover, if $\theta \in (0, 1/2)$, then the corresponding limit does not exist in general. However, in this situation the existence of the limit is always guaranteed at least on the complement of some set of integers of zero density.

Proof. The statements in the theorem concerning partitions α of finite type follow easily from Theorem 3.2.9. Indeed, given that

$$\lim_{n \to \infty} \left(\frac{\lambda\left(C_n^{(\alpha)}\right)}{\left(\sum_{k=1}^{n} t_k\right)^{-1}}\right) = \lim_{n \to \infty} \lambda\left(C_n^{(\alpha)}\right) \cdot \lim_{n \to \infty} \sum_{k=1}^{n} t_k = 1,$$

the statement in part (b) follows immediately. The corresponding claim in part (a) follows directly on considering the Cesàro average of the sequence of Lebesgue measures of α-sum-level sets. Similarly to the proof of Theorem 3.2.9, the remainder of the proof (that is, those parts concerning partitions that are expansive of exponent θ), follow from straightforward applications of the strong renewal results of Garsia/Lamperti and Erickson to the setting of the α-sum-level sets. For this we must set $v_n := a_n$, $V_n := t_n$ and $w_n := \lambda(C_n^{(\alpha)})$, and recall that the so-defined pair of sequences $((v_n)_{n \geq 1}, (w_n)_{n \geq 0})$ satisfies the conditions of a renewal pair. $\qquad\square$

3.3 Exercises

Exercise 3.3.1. Let (v_n) be a infinite probability vector with generating function V (see Definition 3.2.7). Show that

$$\lim_{s \to 1^-} \frac{1 - V(z)}{1 - z} = \sum_{m=1}^{\infty} m \cdot v_m.$$

Exercise 3.3.2. Consider the renewal pair (v_n, w_n) with the extra assumption that $v_1 := p \in (0, 1)$ and $v_2 := 1 - p$. Determine the values of the sequence (w_n) explicitly with the help of the generating functions V and W, and verify that the speed of convergence in the renewal theorem is in fact exponential. (Hint: Use the relation (3.5) to show that $W = 1/(1 - V)$ is a rational function with two poles, one in $s_1 = 1$ and another one in s_2. Find the residues and determine the power series of W by using the identity $(1 - s/s_k) = \sum_{n \geq 0} (s/s_k)^n$, for $k = 1, 2$.)

Exercise 3.3.3. Generalise the ideas of Exercise 3.3.2 to the case that more than two, but only finitely many, of the v_n are non-zero and such that $d_v = 1$. (Hint: As an intermediate step show that $1 - V(s)$ is a polynomial of finite degree which has a simple root in 1 and all the other roots are of modulus strictly greater than 1. Also make use of Exercise 3.3.1.)

Exercise 3.3.4. In the following exercise we employ a very useful representation for slowly varying functions. Assume that L is a slowly varying function. Then there exist constants $c \in \mathbb{R}$ and $A > 0$, a bounded measurable function η and a continuous function δ, both defined on $[a, \infty)$, with $\lim_{x \to \infty} \eta(x) = c$ and $\lim_{x \to \infty} \delta(x) = 0$ such that for all $x \geq a$ we have

$$L(x) = \exp\left(\eta(x) + \int_a^x \frac{\delta(t)}{t} dt\right).$$

Use this representation to prove:
1. For every $0 < r < s < \infty$,

$$\lim_{x \to \infty} \sup_{t \in [r,s]} \left| \frac{L(tx)}{L(x)} - 1 \right| = 0.$$

2. We have

$$\lim_{x\to\infty} \frac{\log(L(x))}{\log(x)} = 0.$$

3. We have for $\alpha > 0$

$$\lim_{x\to\infty} x^\alpha L(x) = \infty \text{ and } \lim_{x\to\infty} x^{-\alpha} L(x) = 0.$$

4. For $\alpha \in \mathbb{R}$ and for slowly varying functions L_1, L_2 and L_3, where also $\lim_{x\to\infty} L_3(x) = \infty$ we have that each of the functions $x \mapsto (L_1(x))^\alpha$, $L_1 + L_2$, and $L_1 \circ L_3$ are also slowly varying.

Exercise 3.3.5. Show that a measurable function $f : \mathbb{R}^+ \to \mathbb{R}^+$ is regularly varying with exponent $\rho \in \mathbb{R}$ if and only if

$$\lim_{x\to\infty} \frac{f(xy)}{f(x)} = y^\rho \text{ for every } y > 0.$$

Exercise 3.3.6. Let us fix a constant $\rho \geq 0$, a slowly varying function ψ and a family of sequences of distribution functions $t \mapsto U^x(t)$, $t \geq 0$, $x \in X$ such that the corresponding measure – also denoted by U^x – carries no mass in 0. The *Laplace transform* of U^x is defined to be

$$\omega^x : \mathbb{R}^+ \to \mathbb{R}^+, s \mapsto \int_0^\infty \exp(-st) \, \mathrm{d}U^x(t).$$

Suppose that uniformly in $x \in X$ we have

$$U^x(t) \sim t^\rho \psi(t),$$

as t tends to infinity, then we also have uniformly in $x \in X$, as τ tends to zero,

$$\omega^x(\tau) \sim \Gamma(\rho + 1)\tau^{-\rho}\psi(1/\tau).$$

Hint: First show that for some $\delta \in (0, 1)$ the quotient $\omega^x(\delta \cdot \tau)/\tau^{-\rho}\psi(1/\tau)$ stays uniformly bounded as τ tends to zero (for this split the domain of integration into the points $2^k/\tau$, $k \in \mathbb{N}$ and use integration by parts). Then use Exercise 3.3.4 (1) with $a > 0$ sufficiently small and $b < \infty$ sufficiently large to split the domain of integration in the definition of the Laplace transform in a convergent part and two negligible parts.

Exercise 3.3.7. Use Exercise 3.3.6 to prove the uniformity statement in Karamata's Tauberian Theorem 3.2.6.
Hint: Consider the distribution function $U^x(t) := \sum_{k=0}^{\lfloor t\rfloor-1} b_k^x + (t - \lfloor t\rfloor)b_{\lfloor t\rfloor}^x$ and make the change of variables $y = \exp(-t)$.

4 Infinite ergodic theory

In this chapter we will make a deeper journey into infinite ergodic theory, taking up where we left off in Chapter 2. In the following chapter, we shall then see some applications of this general theory to continued fractions.

4.1 The functional analytic perspective and the Chacon–Ornstein Ergodic Theorem

Our first main result of this section will be the Chacon–Ornstein Ergodic Theorem [CO60], which is stated completely in terms of linear operators acting on $L_1(\mu)$. After having proved this powerful result, we will then see that it implies Birkhoff's Pointwise Ergodic Theorem (which we have already seen in Theorem 2.4.16), Hopf's Ergodic Theorem (cf. Theorem 2.4.24), as well as Hurewicz's Ergodic Theorem (Corollary 4.1.18), which we will specifically need later in this chapter. The proof, as you might expect, is not trivial. Before stating and proving the Chacon–Ornstein Ergodic Theorem, we will first collect a few useful observations concerning the functional analytic nature of this part of the theory. In particular, we shall now study the previously-defined transfer operator $\widehat{T}$ and Koopman operator $U_T : f \mapsto f \circ T$ from a broader functional-analytic perspective (recall that the Koopman operator U_T was first mentioned in Remark 2.3.8).

Let $(X, \mathcal{B}, \mu)$ denote a σ-finite measure space. We remind the reader that the space of integrable functions $L_1(\mu)$, in which functions are identified if they are a.e. equal, together with norm $\|\cdot\|_1$ given by

$$\|f\|_1 := \int |f| \, d\mu$$

defines a Banach space. The space $L_\infty(\mu)$ equipped with the norm $\|\cdot\|_\infty$ given by

$$\|f\|_\infty := \inf\{c \in \mathbb{R} : |f| \le c \text{ a.e.}\}$$

also defines a Banach space. Further, let $L_p^+(\mu)$ denote the set of non-negative functions from $L_p(\mu)$ and recall that the non-negative part ψ^+ of a measurable real-valued function ψ is given by the measurable function $\psi^+ := \max\{\psi, 0\}$.

We shall now study bounded linear operators acting on $L_1(\mu)$; these are linear functions $V: L_1(\mu) \to L_1(\mu)$ with bounded *operator norm*, which is defined to be

$$\|V\| := \sup_{\|f\|_1 = 1} \|V(f)\|_1 .$$

Definition 4.1.1. Let $V: L_1(\mu) \to L_1(\mu)$ be a linear operator.
(a) V is said to be a *contraction* if $\|V\| \le 1$.
(b) V is said to be *positive* if $V\left(L_1^+(\mu)\right) \subset L_1^+(\mu)$.

We have already seen that if $(X, \mathcal{B}, \mu, T)$ is a non-singular dynamical system then $\widehat{T}: L_1(\mu) \to L_1(\mu)$ is well defined and $\|\widehat{T}\| \le 1$ (see Exercise 2.6.11). In the next lemma, we will consider positivity for $\widehat{T}$ and both properties for the Koopman operator.

Lemma 4.1.2. *For the Koopman and the transfer operator we have:*
(a) *If $(X, \mathcal{B}, \mu, T)$ is a measure-preserving dynamical system, then $U_T : f \mapsto f \circ T$ is a positive contraction on $L_1(\mu)$.*
(b) *If $(X, \mathcal{B}, \mu, T)$ is a non-singular dynamical system, then $\widehat{T}$ is a positive contraction on $L_1(\mu)$.*

Proof. Let $(X, \mathcal{B}, \mu, T)$ be a measure-preserving dynamical system. Then for every $f \in L_1(\mu)$, we have

$$\int |U_T f| \, d\mu = \int |f \circ T| \, d\mu = \int |f| \, d\mu \circ T^{-1} = \int |f| \, d\mu < \infty,$$

which shows that U_T is well defined and a contraction. It is clear that U_T is positive, so the proof of part (a) is finished.

Towards part (b), as we recalled above, we have already seen that $\widehat{T}$ has norm 1. Positivity follows from the fact that for a given $f \in L_1^+(\mu)$ we have, for all $g \in L_\infty^+(\mu)$, that

$$\int \widehat{T} f \cdot g \, d\mu = \int f \cdot g \circ T \, d\mu \ge 0.$$ □

Definition 4.1.3. Let $V: L_1(\mu) \to L_1(\mu)$ be a bounded linear operator. Then the *dual* of V, which will be denoted by V^*, is an operator acting on the dual space $(L_1(\mu))^* \simeq L_\infty(\mu)$ which is uniquely determined by the identity

$$\int f \cdot V^*(g) \, d\mu := \int V(f) \cdot g \, d\mu,$$

for all $f \in L_1(\mu)$ and $g \in L_\infty(\mu)$.

In particular, as was already alluded to in Remark 2.3.8, if $(X, \mathcal{B}, \mu, T)$ is a non-singular dynamical system, then for the transfer operator $\widehat{T}$ we have by definition that $\widehat{T}^* = U_T$ as an operator acting on $L_\infty(\mu)$.

Note that if V is a positive linear operator acting on $L_1(\mu)$ then the dual operator $V^*: L_\infty(\mu) \to L_\infty(\mu)$ is also positive. Indeed, for every $g \in L_\infty^+(\mu)$ we have that $\int V^*(g) \cdot f \, d\mu = \int g \cdot Vf \, d\mu \ge 0$ for all $f \in L_1^+(\mu)$, and hence $V^* g \ge 0$. If V is a positive linear operator acting on $L_1(\mu)$ we will use the notation

$$S_n f := \sum_{k=0}^{n-1} V^k f \text{ for } n \in \mathbb{N} \cup \{\infty\}, \text{ and we set } S_0 f := 0.$$

Note that for $V = U_T$, this definition coincides with our definition in the context of dynamical systems as introduced in Section 2.4.6.

We can now continue towards the proof of the Chacon–Ornstein Ergodic Theorem and its immediate corollaries, i.e., to Hopf's, Birkhoff's and Hurewicz's Pointwise Ergodic Theorems.

Lemma 4.1.4 (Chacon–Ornstein Lemma). *Let V be a positive contraction acting on $L_1(\mu)$ and let $g \in L_1(\mu)$ be such that $g > 0$. For all $\varphi \in L_1(\mu)$, we then have μ-a.e. that*

$$\lim_{n\to\infty} \frac{V^n\varphi}{S_n g} = 0.$$

Proof. Let $\varepsilon > 0$ be fixed and define

$$E_n := \{x \in X : V^n\varphi(x) > \varepsilon S_n g\}.$$

The aim is to show that $\sum_{n=2}^{\infty} \mu_g(E_n) < \infty$, for μ_g given by $d\mu_g := g\,d\mu$, where we assume without loss of generality that $\int g\,d\mu = 1$. This will be sufficient, since then by the Borel–Cantelli Lemma (see Lemma 1.2.18), we have that the set of points which lie in infinitely many of the sets E_n is of μ_g-measure equal to zero; then taking the complement of this limsup set and noting that $g > 0$ as well as that $\varepsilon > 0$ was arbitrary, this gives $\limsup_{n\to\infty} \frac{V^n\varphi}{S_n g} \le 0$ μ-a.e. Applying this result to $-\varphi$ instead of φ shows that μ-a.e. we also have $\liminf_{n\to\infty} \frac{V^n\varphi}{S_n g} \ge 0$, giving finally the assertion.

Now, since V is a positive operator and both 0 and $\psi \in L_1(\mu)$ are less than or equal to ψ^+, we have $0 = V(0) \le V(\psi^+)$ as well as $V(\psi) \le V(\psi^+)$, and hence $(V\psi)^+ \le (V(\psi^+))^+ = V(\psi^+)$. Using this and the fact that $V(V^n\varphi - \varepsilon S_n g) = V^{n+1}\varphi - S_{n+1}\varepsilon g + \varepsilon g$, it follows that

$$\left(V^{n+1}\varphi - \varepsilon S_{n+1}g\right)^+ + \mathbb{1}_{E_{n+1}}\varepsilon g = \mathbb{1}_{E_{n+1}}\left(V^{n+1}\varphi - \varepsilon S_{n+1}g\right) + \mathbb{1}_{E_{n+1}}\varepsilon g$$

$$= \mathbb{1}_{E_{n+1}}\left(V^{n+1}\varphi - S_{n+1}\varepsilon g + \varepsilon g\right)$$

$$= \mathbb{1}_{E_{n+1}}\left(V(V^n\varphi - \varepsilon S_n g)\right)^+$$

$$\le V\left(\left(V^n\varphi - \varepsilon S_n g\right)^+\right).$$

To shorten the notation below, let us set $J_n := \left(V^n\varphi - \varepsilon S_n g\right)^+$. Then the above inequality together with the fact that V is a contraction implies that

$$\varepsilon \int \mathbb{1}_{E_{n+1}} g\,d\mu \le \int (VJ_n - J_{n+1})\,d\mu \le \int (\|V\|J_n - J_{n+1})\,d\mu \le \int (J_n - J_{n+1})\,d\mu.$$

This shows that for all $N \in \mathbb{N}$, we have that

$$\varepsilon \int \sum_{n=1}^{N} \mathbb{1}_{E_{n+1}} g\,d\mu \le \int \sum_{n=1}^{N} (J_n - J_{n+1})\,d\mu = \int J_1 - J_{N+1}\,d\mu \le \int J_1\,d\mu < \infty.$$

$\square$

Lemma 4.1.5 (Hopf's maximal inequality). *Let V be a positive contraction acting on $L_1(\mu)$ and let $f \in L_1(\mu)$. If we set $f_n := \max_{0 \leq j \leq n} S_j f$ for each $n \in \mathbb{N}_0$, we then have*

$$\int_{\{f_n > 0\}} f \, d\mu \geq 0.$$

Proof. First, notice that $f_0 = 0 \leq f^+ = f_1 \leq f_2 \leq \cdots$. Then, using the positivity of V, we have on the set $\{f_n > 0\}$ that

$$f_n = \max_{0 \leq j \leq n} \sum_{k=0}^{j-1} V^k f = \max_{1 \leq j \leq n} \left(f + \sum_{k=1}^{j-1} V^k f \right) = f + \max_{1 \leq j \leq n} V \sum_{k=0}^{j-2} V^k f$$

$$\leq f + V \max_{1 \leq j \leq n} \sum_{k=0}^{j-2} V^k f = f + V(f_{n-1}) \leq f + V(f_n).$$

Since $V f_n \geq 0$ and V is contracting, a rearrangement of the above inequality yields that

$$\int_{\{f_n > 0\}} f \, d\mu \geq \int_{\{f_n > 0\}} (f_n - V f_n) \, d\mu = \int f_n \, d\mu - \int_{\{f_n > 0\}} V f_n \, d\mu$$

$$\geq \int f_n \, d\mu - \int V f_n \, d\mu \geq \int f_n \, d\mu - \|V\| \int f_n \, d\mu \geq 0. \qquad \square$$

Before stating the next result, let us fix some notation. We let $Q_n(\varphi, g) := S_n \varphi / S_n g$ and define $\widetilde{Q}(\varphi, g) := \sup_{n \in \mathbb{N}} Q_n(\varphi, g)$.

Lemma 4.1.6 (Wiener's maximal inequality). *Let V be a positive contraction on $L_1(\mu)$ and let $g \in L_1(\mu)$ be such that $g > 0$ and such that μ_g, given by $d\mu_g := g \, d\mu$, is a probability measure. For each $\varphi \in L_1(\mu)$ and $s > 0$, we then have that*

$$\mu_g \left(\left\{ \widetilde{Q}(\varphi, g) > s \right\} \right) \leq \frac{\|\varphi\|_1}{s}.$$

Proof. Hopf's maximal inequality applied to $f_n := \max_{0 \leq j \leq n} S_j(\varphi - sg)$ gives that for each $n \in \mathbb{N}$, we have that

$$0 \leq \int_{\{f_n > 0\}} (\varphi - sg) \, d\mu = \int_{\{f_n > 0\}} \varphi \, d\mu - s \mu_g(\{f_n > 0\}).$$

Since $f_n > 0$ if and only if $\max_{1 \leq k \leq n} Q_k(\varphi - sg, g) > 0$, it follows that

$$\mu_g \left(\left\{ \max_{1 \leq k \leq n} Q_k(\varphi - sg, g) > 0 \right\} \right) = \mu_g(\{f_n > 0\}) \leq \frac{1}{s} \int_{\{f_n > 0\}} \varphi \, d\mu \leq \frac{1}{s} \|\varphi\|_1.$$

Since $(\{\max_{1\le k\le n} Q_k(\varphi - sg, g) > 0\})_{n\in\mathbb{N}}$ is an increasing sequence of sets with union equal to $\{\tilde{Q}(\varphi, g) > s\}$, using the continuity of μ_g from below finishes the proof. $\square$

Let $g \in L_1(\mu)$ such that $g > 0$. By letting s tend to infinity in the previous lemma, we find that $\tilde{Q}(\varphi, g) < \infty$ μ-a.e., for each $\varphi \in L_1(\mu)$, and in particular for those φ such that $\varphi > 0$. In fact, this shows that $\{S_\infty g = \infty\} = \{S_\infty \varphi = \infty\}$ $\mathrm{mod}\ \mu$ and hence the set $\{S_\infty g = \infty\}$ is μ-a.e. independent of g.

Definition 4.1.7 (Hopf decomposition for operators). Let V be a positive contraction on $L_1(\mu)$. The above observation allows us to define the (μ-a.e. determined) *Hopf decomposition* of X with respect to the positive contraction V into the *conservative part* $C_V := \{S_\infty g = \infty\}$ for some $g \in L_1(\mu)$ such that $g > 0$ and the *dissipative part* $D_V := X\backslash C_V$. If $X = C_V$ $\mathrm{mod}\ \mu$, then V is called *conservative*.

Remark 4.1.8. This definition further generalizes our notion of Hopf decompositions. Let $(X, \mathcal{B}, \mu, T)$ be a measure-theoretical dynamical system. Then the following hold:
(a) If the system is measure-preserving, then
$$C_T = C_{U_T} \quad \mathrm{mod}\ \mu.$$

(b) If the system is non-singular, then
$$\widehat{C}_T = C_{\widehat{T}} \quad \mathrm{mod}\ \mu.$$

Lemma 4.1.9. *Let V be a conservative positive contraction on $L_1(\mu)$ and let $\varphi \in L_\infty(\mu)$ be given such that either $V^*\varphi \ge \varphi$ or $V^*\varphi \le \varphi$. Then $V^*\varphi = \varphi$.*

Proof. For the case that $V^*\varphi \ge \varphi$, let $g \in L_1(\mu)$ be fixed such that $g > 0$. We then have that
$$0 \le \int \left(V^*\varphi - \varphi\right) \sum_{k=0}^{n-1} V^k g\, d\mu = \int \varphi\, (V^n g - g)\, d\mu \le 2\,\|\varphi\|_\infty \int g\, d\mu < \infty.$$

Since by assumption $X_\infty(g) = X$, we have that $S_n g = \sum_{k=0}^{n-1} V^k g$ is unbounded μ-a.e. on X. Therefore, the latter inequality can be satisfied only if $V^*\varphi = \varphi$.
 The case $V^*\varphi \le \varphi$ can be treated in an analogous way and is left to the reader. $\square$

Example 4.1.10. Since for all $f \in L_1^+(\mu)$,
$$\int f \cdot V^* \mathbb{1}_X\, d\mu = \int Vf\, d\mu \le \int f\, d\mu = \int f \cdot \mathbb{1}_X\, d\mu,$$

it follows that $V^*\mathbb{1}_X \le \mathbb{1}_X$. Consequently, in light of Lemma 4.1.9, we deduce that $V^*\mathbb{1}_X = \mathbb{1}_X$.

Lemma 4.1.11. *Let V be a positive conservative contraction on $L_1(\mu)$ and let $\varphi \in L_\infty(\mu)$ be V^*-invariant, that is, $\varphi = V^*\varphi$. Then φ^+ and $\mathbb{1}_{\{a<\varphi\leq b\}}$ are also both V^*-invariant, for all $a,b \in \mathbb{R}$.*

Proof. Fix $\varphi \in L_\infty(\mu)$ such that $\varphi = V^*\varphi$. Since V^* is a positive operator, we have that $V^*\varphi^+ \geq (V^*\varphi)^+ = \varphi^+$ and hence we can apply Lemma 4.1.9, which tells us that $V^*\varphi^+ = \varphi^+$. These observations together with Example 4.1.10 give that for each $a \in \mathbb{R}$,

$$V^*(\varphi - a\mathbb{1}_X) = \varphi - a\mathbb{1}_X \text{ and } V^*(\varphi - a\mathbb{1}_X)^+ = (\varphi - a\mathbb{1}_X)^+.$$

Next, observe that the sequence $\left(h_n := n\left(1/n - (1/n - (\varphi - a)^+)^+\right)\right)_{n\geq 1}$ converges monotonically from below to the indicator function $\mathbb{1}_{\{a<\varphi\}}$. Since by the above observations, all elements h_n in this sequence are V^*-invariant, we get $V^*\mathbb{1}_{\{a<\varphi\}} \geq V^*h_n = h_n \nearrow \mathbb{1}_{\{a<\varphi\}}$. Again by Lemma 4.1.9 we have that $V^*\mathbb{1}_{\{a<\varphi\}} = \mathbb{1}_{\{a<\varphi\}}$. Now, the lemma follows on observing that $\mathbb{1}_{\{a<\varphi\leq b\}} = \mathbb{1}_{\{a<\varphi\}} - \mathbb{1}_{\{b<\varphi\}}$. □

For a positive conservative contraction, the fixed points of its dual can be characterized in the following way.

Lemma 4.1.12. *Let V be a positive conservative contraction on $L_1(\mu)$ and let $\varphi \in L_\infty(\mu)$. Then the following equivalence holds:*

$$V^*\varphi = \varphi \iff V(\varphi \cdot h) = \varphi \cdot V(h), \text{ for all } h \in L_1^+(\mu).$$

Proof. Let $\varphi \in L_\infty(\mu)$ and assume that we have $V(\varphi \cdot h) = \varphi \cdot V(h)$, for all $h \in L_1^+(\mu)$. Note that by Example 4.1.10, we have that $\mathbb{1}_X = V^*\mathbb{1}_X$. Using this, it follows that for all $h \in L_1^+(\mu)$ we have

$$\int V^*(\varphi) \cdot h \, d\mu = \int \varphi \cdot V(h) \, d\mu = \int V(\varphi \cdot h) \, d\mu$$
$$= \int V^*(\mathbb{1}_X) \cdot \varphi \cdot h \, d\mu = \int \varphi \cdot h \, d\mu.$$

This shows that $\varphi = V^*\varphi$.

For the reverse direction, let $\varphi \in L_\infty(\mu)$ be given such that $V^*\varphi = \varphi$. By Lemma 4.1.11, we have for $F := \{a < \varphi \leq b\}$, for arbitrary $a,b \in \mathbb{R}$, that $\mathbb{1}_F$ and $\mathbb{1}_{F^c} = \mathbb{1}_X - \mathbb{1}_F$ are both V^*-invariant. Using this, it follows that for each $h \in L_1^+(\mu)$ we have that

$$0 = \int \underbrace{\mathbb{1}_{F^c} \cdot V(\mathbb{1}_F h)}_{\geq 0} \, d\mu = \int \mathbb{1}_F V(\mathbb{1}_{F^c} h) \, d\mu.$$

It follows that $\mathbb{1}_{F^c} V(\mathbb{1}_F h) = \mathbb{1}_F V(\mathbb{1}_{F^c} h) = 0$. Using this and the linearity of V, we obtain

$$V(\mathbb{1}_F h) = (\mathbb{1}_F + \mathbb{1}_{F^c}) V(\mathbb{1}_F h) = \mathbb{1}_F(V(h) - V(\mathbb{1}_{F^c} h)) = \mathbb{1}_F V(h).$$

Now the claim follows by approximating φ in $L_\infty(\mu)$-norm by elementary functions of the form $\varphi_n := \sum_{k=-2^n N}^{2^n N} 2^{-n} k \mathbb{1}_{\{k2^{-n} < \varphi \le (k+1)2^{-n}\}}$, for some fixed $N > \|\varphi\|_\infty$. Note that we have $\|\varphi_n h - \varphi h\|_1 \to 0$, for $n \to \infty$. $\qquad\square$

Remark 4.1.13. The above lemma shows in particular that if all $\varphi_k \in L_\infty(\mu)$, $k \in \{1, \dots, n\}$, are V^*-invariant then so is their product $\varphi_1 \cdots \varphi_n$.

Definition 4.1.14. Let $V: L_1(\mu) \to L_1(\mu)$ be a bounded linear operator. Then the system $(L_1(\mu), V)$ is called *ergodic* if the σ-algebra $\mathcal{I} := \sigma(\{f \in L_\infty(\mu): V^* f = f\})$ generated by the V^*-invariant functions is trivial.

Remark 4.1.15. Note that the σ-algebra $\mathcal{I}$ is trivial if and only if $g \in \{f \in L_\infty(\mu): V^* f = f\}$ implies that g is constant.

Lemma 4.1.16. *For the transfer and the Koopman operator we have:*
(a) *If $(X, \mathcal{B}, \mu, T)$ is a measure-preserving, conservative and ergodic dynamical system, then $(L_1(\mu), U_T)$ is conservative and ergodic.*
(b) *If $(X, \mathcal{B}, \mu, T)$ is a non-singular, conservative and ergodic dynamical system, then $(L_1(\mu), \widehat{T})$ is conservative and ergodic.*

Proof. Both systems $(L_1(\mu), U_T)$ and $(L_1(\mu), \widehat{T})$ are conservative by Remark 4.1.8. In order to prove the ergodicity of U_T, we have to show that the σ-algebra $\mathcal{I}$ generated by the U_T^*-invariant functions is trivial, which is equivalent to the fact that all U_T^*-invariant functions are constant. Fix $\varphi \in L_\infty(\mu)$ such that $U_T^* \varphi = \varphi$. Then, by Lemma 4.1.11, we can assume without loss of generality that $\varphi \in L_\infty^+(\mu) \backslash \{0\}$. Therefore, where μ_φ is given by $d\mu_\varphi := \varphi d\mu$, we have for all $f \in L_1(\mu)$ that

$$\int f \circ T \, d\mu_\varphi = \int U_T f \cdot \varphi \, d\mu = \int f U_T^*(\varphi) \, d\mu = \int f\varphi \, d\mu = \int f \, d\mu_\varphi.$$

This shows that μ_φ is an invariant ergodic measure absolutely continuous to μ. Hence, Theorem 2.4.35 implies that μ_φ is equal to $c \cdot \mu$ for some constant $c > 0$. This shows that $\varphi = c$ is in fact a constant function.

For part (b), to prove that $\widehat{T}$ is ergodic, we must show that the σ-algebra $\mathcal{I}$ generated by the $\widehat{T}^*$-invariant functions is trivial. However, since $\widehat{T}^*$ coincides with the Koopman operator U_T acting on $L_\infty(\mu)$, the latter assertion is an immediate consequence of Proposition 2.4.6. $\qquad\square$

We are now in a position to state and prove our first main result of this chapter. Before stating the theorem, we recall that for every sub-σ-algebra $\mathcal{F}$ of a σ-algebra $\mathcal{B}$, and function $g \in L_1(\mu)$ there exists an a.s. uniquely-defined function $\mathbb{E}_\mu(g|\mathcal{F})$ called the *conditional expectation of g with respect to $\mathcal{F}$* which can be characterised as follows: $f = \mathbb{E}_\mu(g|\mathcal{F})$ if and only if f is $\mathcal{F}$-measurable and for all $B \in \mathcal{F}$ we have

$$\int_B f \, d\mu = \int_B g \, d\mu.$$

Note that the conditional expectation is already characterised if the above equality can be shown to hold for sets $B \in \mathcal{F}'$ with $\mathcal{F} = \sigma(\mathcal{F}')$ and such that $\mathcal{F}'$ is closed under intersections and contains Ω.

Theorem 4.1.17 (Chacon–Orstein Ergodic Theorem). *Let V be a positive conservative contraction on $L_1(\mu)$. We then have for $f, g \in L_1(\mu)$ such that $g > 0$, that the sequence of quotients $Q_n(f, g) := S_n f / S_n g$ converges μ-a.e. to a function $Q(f, g)$, for n tending to infinity. Moreover, the function $Q(f, g)$ can be expressed as a conditional expectation. That is,*

$$Q(f, g) = \mathbb{E}_{\mu_g}(f/g \mid \mathcal{I}),$$

where $\mathcal{I}$ denotes σ-algebra generated by the V^*-invariant functions and μ_g is given by $d\mu_g := g/\mu(g)\, d\mu$.
Furthermore, if the system $(L_1(\mu), V)$ is ergodic (that is, $\mathcal{I}$ is trivial), then the limiting function $Q(f, g)$ is μ-a.e. equal to the constant function $\int f\, d\mu / \int g\, d\mu$.

Proof. Let us begin by proving the μ-a.e. convergence. For this, fix $g \in L_1(\mu)$ such that $g > 0$. Next, let us define the set

$$L := \left\{ \varphi g + \psi - V\psi : \varphi \in L_\infty(\mu), \psi \in L_1(\mu), V(\varphi h) = \varphi V(h), \text{ for all } h \in L_1^+(\mu) \right\}.$$

Now, let $f \in L$ be fixed. By the definition of L, we immediately obtain that

$$Q_n(f, g) = \frac{\varphi S_n g + (\psi - V^n \psi)}{S_n g}, \quad \text{for all } n \in \mathbb{N}.$$

Recalling that V is conservative, a straightforward application of the Chacon–Ornstein Lemma shows that on $X_\infty(g) = X$ the sequence $(Q_n(f, g))_{n \geq 1}$ converges μ-a.e. to the function φ. Therefore, we have that

$$L \subset \mathcal{L} := \{ h \in L_1(\mu) : (Q_n(h, g))_{n \geq 1} \text{ converges } \mu\text{-a.e.} \}$$

Our next step is to show that L is a dense subset of $L_1(\mu)$ and that $\mathcal{L}$ is closed. To show the denseness, we use the general fact that a subspace is dense in a normed vector space if and only if the annihilator of the subspace is equal to the null-space, which is a consequence of the Hahn–Banach Theorem (cf. [Rud91, Theorem 4.7]). Hence, it is sufficient to show that the following implication holds for each $k \in L_\infty(\mu)$:

$$\int kh\, d\mu = 0, \text{ for all } h \in L \quad \Longrightarrow \quad k = 0. \tag{4.1}$$

In order to show this, note that for each $\psi \in L_1(\mu)$, we have that $\psi - V\psi$ belongs to L. Hence, for each $k \in L_\infty(\mu)$ that fulfills the left hand side of the above implication, we have

$$0 = \int k\psi \, d\mu - \int kV\psi \, d\mu = \int (k - V^*k)\psi \, d\mu.$$

Since this must hold for all $\psi \in L_1(\mu)$, it follows that $k = V^*k$. Using Lemma 4.1.12, we have that $kg \in L$ and hence, $\int kkg \, d\mu = 0$. Since $kkg \geq 0$, it follows that $k = 0$, which proves (4.1).

Now, to establish the μ-a.e. convergence, all that is left to show is that $\mathcal{L}$ is closed in $L_1(\mu)$. For this, let us fix $h \in \overline{\mathcal{L}}$ and $\delta > 0$ such that for each $\varepsilon > 0$ there exists $h_\varepsilon \in \mathcal{L}$ with $\|h - h_\varepsilon\|_1 < \varepsilon \cdot \delta$. Since we have that

$$\limsup_{n,m \to \infty} |Q_n(h,g)(x) - Q_m(h,g)(x)|$$

$$= \limsup_{n,m \to \infty} |Q_n(h - h_\varepsilon, g)(x) - Q_m(h - h_\varepsilon, g)(x)|,$$

an application of Wiener's maximal inequality (Lemma 4.1.6) gives that

$$\mu_g\left(\left\{x : \limsup_{n,m \to \infty} |Q_n(h,g)(x) - Q_m(h,g)(x)| > \delta\right\}\right)$$

$$\leq \mu_g\left(\left\{x : \max_{n,m} |Q_n(h - h_\varepsilon, g)(x) - Q_m(h - h_\varepsilon, g)(x)| > \delta\right\}\right)$$

$$\leq \mu_g\left(\left\{x : \widetilde{Q}(|h - h_\varepsilon|, g)(x) > \delta/2\right\}\right) \leq 2\frac{\|h - h_\varepsilon\|_1}{\delta} \leq 2\varepsilon.$$

Since $\delta > 0$ was chosen arbitrarily, we can now conclude, by letting ε tend to zero, that μ_g-a.e. we have that $(Q_n(h,g))_{n \geq 1}$ converges. This shows that $h \in \mathcal{L}$ and hence, $\mathcal{L}$ is closed.

It remains to characterise the limiting function. In order to do so, let $\mathbb{E}(\cdot | \mathcal{I}) = \mathbb{E}_{\mu_g}(\cdot | \mathcal{I})$ denote the conditional expectation with respect to the measure μ_g. Let us first consider the special situation in which $f = \varphi g + \psi - V\psi$ is a fixed element in L. We already know that in this case the limit of $(Q_n(f,g))_{n \geq 1}$ is μ-a.e. equal to φ, which is $\mathcal{I}$-measurable, by Lemma 4.1.12. Let $\mathcal{J}$ denote the set of all finite intersections of subsets of the form $\{x : a < h(x) \leq b\}$, for some $a, b \in \mathbb{R}$ and $h \in L_\infty(\mu)$ with $h = V^*h$. Then using Lemma 4.1.11 and Remark 4.1.13, we have for all $F \in \mathcal{J}$,

$$\mu(g) \int \mathbb{1}_F \cdot f/g \, d\mu_g = \int \mathbb{1}_F \cdot f \, d\mu = \int \mathbb{1}_F \varphi g \, d\mu + \int \mathbb{1}_F \cdot (\psi - V(\psi)) \, d\mu$$

$$= \int \mathbb{1}_F \cdot \varphi g \, d\mu + \int \mathbb{1}_F \cdot \psi \, d\mu - \int V^*(\mathbb{1}_F) \cdot \psi \, d\mu$$

$$= \mu(g) \int \mathbb{1}_F \cdot \varphi \, d\mu_g.$$

Since the set $\mathcal{J}$ is closed under taking intersections and generates $\mathcal{I}$ and since φ is $\mathcal{I}$-measurable, it follows from the characterisation of the conditional expectation stated above that $\varphi = \mathbb{E}\left(f/g|\mathcal{I}\right)$. For general $f \in L_1(\mu)$ the claim follows by approximating f with functions from L.

Finally, if we additionally have that V is ergodic, then the σ-algebra $\mathcal{I}$ generated by the V^*-invariant functions is trivial by definition and hence, the limiting function is μ-a.e. constant and equal to

$$\mathbb{E}\left(f/g|\mathcal{I}\right) = \int f/g \, d\mu_g = \frac{\int f \, d\mu}{\int g \, d\mu}.$$

$\square$

As already mentioned at the beginning, we end this section by first showing how the Chacon–Ornstein Ergodic Theorem implies Hurewicz's Ergodic Theorem, and then Hopf's Ergodic Theorem, and, in turn, Birkhoff's Pointwise Ergodic Theorem.

Corollary 4.1.18 (Hurewicz's Ergodic Theorem). *Let $(X, \mathcal{B}, \mu, T)$ be an ergodic conservative dynamical system and let $g \in L_1^+(\mu)$ be such that $\int g \, d\mu > 0$. For each $f \in L_1(\mu)$, we then have μ-a.e. that*

$$\lim_{n \to \infty} \frac{\sum_{k=0}^{n-1} \widehat{T}^k(f)}{\sum_{k=0}^{n-1} \widehat{T}^k(g)} = \frac{\int f \, d\mu}{\int g \, d\mu}.$$

Proof. By the observations in Section 4.1 for $V = \widehat{T} \colon L_1(\mu) \to L_1(\mu)$, the Chacon–Ornstein Ergodic Theorem is applicable. Hence, for all $f \in L_1(\mu)$ and for a particular $g_0 \in L_1^+(\mu)$ such that $g_0 > 0$ the convergence holds as stated in the corollary. Note that such a function g_0 exists because $(X, \mathcal{B}, \mu)$ is σ-finite; in fact, this is a necessary and sufficient condition for σ-finiteness. For general $g \in L_1^+(\mu)$ with $\int g \, d\mu > 0$ we have that

$$\lim_{n \to \infty} \frac{\sum_{k=0}^{n-1} \widehat{T}^k(f)}{\sum_{k=0}^{n-1} \widehat{T}^k(g)} = \lim_{n \to \infty} \left(\frac{\sum_{k=0}^{n-1} \widehat{T}^k(f)}{\sum_{k=0}^{n-1} \widehat{T}^k(g_0)} \middle/ \frac{\sum_{k=0}^{n-1} \widehat{T}^k(g)}{\sum_{k=0}^{n-1} \widehat{T}^k(g_0)} \right) = \frac{\int f \, d\mu}{\int g \, d\mu}.$$

$\square$

Remark 4.1.19. We can also give a second proof of Hopf's Ergodic Theorem (which we already discussed in Chapter 2, see Theorem 2.4.24), using the Chacon–Orstein Ergodic Theorem. The argument works in precisely the same way as that given above. First, we have seen in Section 4.1 that the Chacon–Ornstein Ergodic Theorem is applicable to the Koopman operator $U_T \colon L_1(\mu) \to L_1(\mu)$, given by $U_T f := f \circ T$. Therefore, the assertion in Hopf's Ergodic Theorem certainly holds for all $f \in L_1(\mu)$ and for a particular $g_0 \in L_1^+(\mu)$ such that $g_0 > 0$. The proof is then completed in exactly the same way as in Hurewicz's Ergodic Theorem for any $g \in L_1(\mu)$ with $\int g \, d\mu > 0$.

Now Birkhoff's Pointwise Ergodic Theorem (see Theorem 2.4.16) follows immediately by choosing $g = \mathbb{1}_X$ in Hopf's Ergodic Theorem. Recall that in the proof we gave of Hopf's Ergodic Theorem in Chapter 2 using inducing, part of the argument was to use Birkhoff's Theorem, so this deduction is only reasonable now.

4.2 Pointwise dual ergodicity

In this section we will study an ergodicity property of the transfer operator associated to a system.

Definition 4.2.1. An ergodic, conservative measure-preserving system $(X, \mathcal{B}, \mu, T)$ is said to be *pointwise dual ergodic* if there exists a sequence $(r_n)_{n \geq 1}$ such that μ-a.e.

$$\frac{1}{r_n} \sum_{i=0}^{n-1} \widehat{T}^i f \to \int f \, d\mu \text{ for all } f \in L_1(\mu).$$

The sequence $(r_n)_{n \geq 1}$, which is unique up to asymptotic equivalence (see remark below), will be referred to as the *return sequence of T*.

Remark 4.2.2. For what follows it will be important to find a sequence $(a_n)_{n \geq 1}$ in the asymptotic class of (r_n) that is strictly increasing. First fix $f, g \in \mathcal{M}(\mathcal{B})$ with f a μ-integrable function and g a bounded function with $g > 0$, $\int g \, d\mu = 1$. Then fix a μ-typical point $x \in X$ that witnesses both the Hurewicz Ergodic Theorem, in the sense that

$$\lim_{n \to \infty} \sum_{k=0}^{n} \widehat{T}^k f(x) / \sum_{k=0}^{n} \widehat{T}^k g(x) = \int f \, d\mu,$$

and the pointwise dual ergodicity for f, in the sense that

$$\frac{1}{r_n} \sum_{i=0}^{n-1} \widehat{T}^i f(x) \to \int f \, d\mu.$$

Then the sequence $a_n := \sum_{k=0}^{n-1} \widehat{T}^k g(x)$ is strictly increasing and asymptotic to r_n. In fact we have shown that a_n can be written as the sum $a_n = \sum_{k=1}^{n} b_k$ where $b_k := \widehat{T}^{k-1} g(x)$ is strictly positive.

Definition 4.2.3. For a set $A \in \mathcal{B}$ with $0 < \mu(A) < \infty$, the *wandering rate* of A with respect to T is given by the sequence $(w_n(A))_{n \geq 1}$, where

$$w_n(A) := \mu \left(\bigcup_{k=0}^{n} T^{-k}(A) \right).$$

Where φ is the return time function with respect to A (as in Definition 2.4.25), we also have that

$$w_n(A) = \sum_{k=0}^{n} \mu(A \cap \{\varphi > k\}) = \mu(A) + \sum_{k=1}^{n} \mu \left(T^{-k} A \setminus \bigcup_{\ell=0}^{k-1} T^{-\ell} A \right). \tag{4.2}$$

The proof of this statement is left to Exercise 4.4.8

Information about the wandering rate can be understood as information about "how infinite" the system is, in terms of the size of X relative to E. That is, the increments $\mu(A \cap \{\varphi > n\})$ in the characterisation of the wandering rate quantify in some way how large X is relative to E.

Definition 4.2.4. Let $(X, \mathcal{B}, \mu, T)$ be pointwise dual ergodic with return sequence (r_n). Then, a set $A \in \mathcal{B}$ with positive, finite measure is called a *uniform set* for $f \in L_1^+(\mu)$ if

$$\frac{1}{r_n} \sum_{k=0}^{n-1} \widehat{T}^k f \to \int f \, d\mu \text{ uniformly (mod } \mu) \text{ on } A.$$

We will also call a set $A \in \mathcal{B}$ uniform if it is a uniform set for some $f \in L_1^+(\mu)$ with $\int f \, d\mu > 0$.

Here, uniform convergence (mod μ) on A means uniform convergence on a set A_0 such that the symmetric difference $A_0 \triangle A$ has μ-measure zero. One can also think of this convergence as convergence in the $L_\infty(\mu|_A)$-norm. We also recall that the definition of a regularly varying sequence was given directly above Theorem 3.2.6.

Lemma 4.2.5. *Assume that the sequence (r_n) is regularly varying with exponent $\rho \in [0, 1]$ and given by $r_n = \sum_{k=0}^n b_k$, for some non-negative sequence (b_n). If for some $A \in \mathcal{B}$ with $0 < \mu(A) < \infty$ and $f \in L_1^+(\mu)$ we have uniformly (mod μ) on A that*

$$\lim_{n \to \infty} \frac{1}{r_n} \sum_{k=0}^{n-1} \widehat{T}^k f = \int f \, d\mu,$$

then with $B(s) := \sum_{n=0}^\infty b_n s^n$, $s \in [0, 1)$, we have uniformly (mod μ) on A that

$$\lim_{s \to 1^-} \frac{1}{B(s)} \sum_{n=0}^\infty s^n \widehat{T}^n f = \int f \, d\mu.$$

Proof. Let $r_n = n^\rho \psi(n)$ for some slowly varying function ψ. Since $b_n \le r_n$ it follows that $B(s)$ is finite for all $s \in [0, 1)$. Then the claim follows by applying Karamata's Tauberian Theorem (see Theorem 3.2.6) twice, first with the sequence (b_n) and then with the sequence $(\widehat{T}^n f)$. More precisely, first note that

$$\frac{1}{\Gamma(1+\rho)} \sum_{k=0}^n b_k \sim \frac{n^\rho \psi(n)}{\Gamma(1+\rho)} \implies B(s) \sim \frac{1}{(1-s)^\rho} \psi\left(\frac{1}{1-s}\right) \Gamma(1+\rho).$$

Thus, since by assumption uniformly (mod μ) on A

$$\sum_{k=0}^n \widehat{T}^k f \sim \int f \, d\mu \cdot r_n \sim \int f \, d\mu \cdot n^\rho \psi(n)$$

it follows that uniformly (mod μ) on A

$$B(s) \cdot \int f \, d\mu \sim \frac{\int f \, d\mu}{(1-s)^\rho} \psi \left(\frac{1}{1-s} \right) \Gamma(1+\rho) \sim \sum_{n=0}^{\infty} s^n \widehat{T}^n f.$$

□

Lemma 4.2.6 ([Aar81]). *For* $f \in L_1(\mu)$ *or* $f \in M^+(\mathcal{B})$, *and* $A \in \mathcal{B}$ *such that* $0 < \mu(A) < \infty$ *we have for* $s \in (0,1)$

$$\int_A (1-s^\varphi) \sum_{n=0}^{\infty} s^n \widehat{T}^n f \, d\mu = \sum_{n=0}^{\infty} s^n \int_{A_n} f \, d\mu,$$

where $A_n := T^{-n}A \setminus \bigcup_{k=0}^{n-1} T^{-k}A$, *for* $n \in \mathbb{N}$, *and* $A_0 = A$.

Proof. We may restrict our attention to non-negative measurable functions f only since the case $f \in L_1(\mu)$ can be deduced from this in the following way: First note that for $f \in L_1^+(\mu)$ the right hand side of the equality is finite due to the fact that the sets A_n are pairwise disjoint. For an arbitrary $f \in L_1(\mu)$ we consider its positive part $\max(f,0)$ and its negative part $\max(-f,0)$ separately and use the linearity and positivity of $\widehat{T}$ together with the linearity of both the integral and the infinite sum.

For $f \in M^+(\mathcal{B})$ and with $d\nu = f \, d\mu$ we have

$$\int_A \widehat{T}^n f \, d\mu = \nu(T^{-n}A)$$

$$= \nu(A_n) + \sum_{k=0}^{n-1} \nu \left(T^{-k}(A \cap \{\varphi = n-k\}) \right)$$

$$= \nu(A_n) + \int_A \sum_{k=0}^{n-1} \widehat{T}^k f \cdot \mathbb{1}_{\{\varphi = n-k\}} \, d\mu.$$

Using this gives

$$\int_A \sum_{n=0}^{\infty} s^n \widehat{T}^n f \, d\mu = \sum_{n=0}^{\infty} s^n \nu(A_n) + \sum_{n=1}^{\infty} s^n \int_A \sum_{k=0}^{n-1} \widehat{T}^k f \cdot \mathbb{1}_{\{\varphi = n-k\}} \, d\mu$$

$$= \sum_{n=0}^{\infty} s^n \nu(A_n) + \int_A \sum_{k=1}^{\infty} s^k \mathbb{1}_{\{\varphi = k\}} \sum_{n=k}^{\infty} s^{n-k} \widehat{T}^{n-k} f \, d\mu$$

$$= \sum_{n=0}^{\infty} s^n \int_{A_n} f \, d\mu + \int_A s^\varphi \sum_{n=0}^{\infty} s^n \widehat{T}^n f \, d\mu.$$

Rearranging the last identity proves the claim. □

Proposition 4.2.7 (Asymptotic Renewal Equation [Aar81]). *Let A be a uniform set with regular varying return sequence* $r_n = \sum_{k=0}^n b_k$, *for some non-negative sequence* (b_k).

Then

$$\int_A (1 - s^\varphi)\, d\mu \sim \frac{1}{B(s)}, \quad \text{for } s \to 1^-,$$

where $B(s) := \sum_{n=0}^\infty b_n s^n$.

Proof. Let A be a uniform set for $f \in L_1^+(\mu)$. By Lemma 4.2.6 and with A_n as defined therein, we have

$$\int_A (1 - s^\varphi) \sum_{n=0}^\infty s^n \widehat{T}^n f\, d\mu = \sum_{n=0}^\infty s^n \int_{A_n} f\, d\mu \to \int f\, d\mu,$$

for $s \to 1^-$, where the convergence follows by Abel's Theorem. On the other hand, making use now of Lemma 4.2.5 and the almost everywhere uniform convergence, we find for $s \to 1^-$

$$\int_A (1 - s^\varphi) \sum_{k=0}^\infty s^k \widehat{T}^k f\, d\mu \sim B(s) \int f\, d\mu \int_A (1 - s^\varphi)\, d\mu.$$

These observations combined prove the proposition. □

Proposition 4.2.8. *If T is pointwise dual ergodic with return sequence (r_n) and if A is a uniform set such that the wandering rate $w_n(A)$ is regularly varying with exponent $\alpha \in [0, 1]$, then we have*

$$r_n\, w_n(A) \sim \frac{n}{\Gamma(2 - \alpha)\Gamma(1 + \alpha)}.$$

In particular, we have that r_n is regularly varying with exponent $1 - \alpha$ and there exists a sequence $W_n \nearrow \infty$ such that for all uniform sets $B \in \mathcal{B}$ we have $W_n \sim w_n(B)$.

Proof. As shown in Remark 4.2.2, pointwise dual ergodicity implies that the sequence (r_n) can be chosen to be $r_n = \sum_{k=0}^n b_n$ for some strictly positive sequence (b_n). On the one hand, Lemma 4.2.6 for $f = \mathbb{1}_X$ together with Proposition 4.2.7 implies, for $s \to 1^-$,

$$\sum_{k=0}^\infty s^k \mu(A_k) = \frac{1}{1 - s} \int_A (1 - s^\varphi)\, d\mu \sim \frac{1}{(1 - s)B(s)},$$

with $B(s) := \sum_{k=0}^\infty s^k b_k$. Since, as in (4.2),

$$w_n(A) = \mu(A) + \sum_{k=1}^n \mu(A_k) \sim n^\alpha \psi(n)$$

for $\alpha \in [0, 1]$ and ψ some slowly varying function, Karamata's Tauberian Theorem gives on the other hand

$$\sum_{k=1}^{\infty} s^k \mu(A_k) \sim \frac{\Gamma(1 + \alpha)}{(1 - s)^{\alpha}} \psi\left(\frac{1}{1 - s}\right).$$

Hence we have

$$B(s) \sim \frac{1}{(1 - s)^{1 - \alpha} \Gamma(1 + \alpha) \psi(1/(1 - s))}.$$

Applying in Karamata's Tauberian Theorem the assertion (a) implies (b) to the generating function $B(s)$ gives

$$r_n = \sum_{k=0}^{n} b_k \sim n^{1-\alpha} \frac{1}{\Gamma(2 - \alpha)\Gamma(1 + \alpha)\psi(n)}.$$

From this the claim follows. $\qquad\square$

4.3 ψ-mixing, Darling–Kac sets and pointwise dual ergodicity

Definition 4.3.1. Let $(X, \mathcal{B}, \mu, T)$ be a measure-preserving system. Then a set $A \in \mathcal{B}$ with positive, finite measure is called a *Darling–Kac set* if there exists a positive sequence (r_n) such that

$$\frac{1}{r_n} \sum_{k=0}^{n-1} \widehat{T}^k \mathbb{1}_A \to \mu(A) \text{ uniformly (mod } \mu \text{) on } A.$$

Let us now introduce a stronger mixing property than the one given in Definition 2.5.8. We recall that the refinements $\mathcal{U}^n$ of a collection of sets $\mathcal{U}$ were defined in Definition 1.2.23(c) and $\sigma(\mathcal{U})$ denotes the σ-algebra generated by $\mathcal{U}$.

Definition 4.3.2. Let $(X, \mathcal{B}, \mu, T)$ be a dynamical system with $\mu(X) < \infty$ and let $\mathbb{M}$ be a measurable partition of X. Then the system is said to be ψ-*mixing* with respect to $\mathbb{M}$, if there exists a sequence $(\psi_m)_{m \geq 0}$ of positive real numbers which tends to zero for m tending to infinity, such that for all $n \in \mathbb{N}$, $A \in \sigma(\mathbb{M}^n)$, $B \in \mathcal{B}$ and $m \in \mathbb{N}_0$, we have that

$$\mu\left(A \cap T^{-(m+n)}(B)\right) \leq (1 + \psi_m)\mu(A)\mu(B),$$

and for all $m \in \mathbb{N}$ large enough

$$\mu\left(A \cap T^{-(m+n)}(B)\right) \geq (1 - \psi_m)\mu(A)\mu(B).$$

Proposition 4.3.3. *Let $(X, \mathcal{B}, \mu, T)$ be a conservative, measure-preserving and ergodic dynamical system. Let $A \in \mathcal{B}$ with $0 < \mu(A) < \infty$ and let $\mathbb{M}$ be a measurable partition of A such that the return time function φ with respect to A is $\mathbb{M}$-measurable and the induced system $(A, \mathcal{B}_A, \mu|_A, T_A)$ is ψ-mixing with respect to $\mathbb{M}$. Then A is a Darling–Kac set.*

Proof. Without loss of generality we assume that $\mu(A) = 1$. Recall that since T is conservative and ergodic we have by Corollary 2.4.5 that $\sum_{k=0}^{\infty} \widehat{T}^k \mathbb{1}_A = \infty$ and hence the sequence $(a_n)_{n \geq 1}$ given for each $n \in \mathbb{N}$ by $a_n := \sum_{k=1}^{n} \mu(A \cap T^{-k}A)$ tends to infinity. We will show that this sequence will witness the Darling–Kac property for A. To begin, by the assumed ψ-mixing condition there exists a positive sequence $(\psi_n)_{n \geq 0}$ with $\lim_{n \to \infty} \psi_n = 0$ such that for $B \in \sigma(\mathbb{M}^k)$ with $k \in \mathbb{N}$, and for all $n \in \mathbb{N}_0$ we have

$$\widehat{T}_A^{k+n} \mathbb{1}_B \leq (1 + \psi_n)\mu(B)$$

and for all $n \in \mathbb{N}$ large enough

$$\widehat{T}_A^{k+n} \mathbb{1}_B \geq (1 - \psi_n)\mu(B).$$

The key observation in this proof will be that for $B \in \mathcal{B}$ with $B \subset A$,

$$A \cap T^{-n}B = \bigcup_{k=1}^{n} \{\varphi_k = n\} \cap T_A^{-k}B, \tag{4.3}$$

where, as in (2.12), we define $\varphi_k := \sum_{\ell=0}^{k-1} \varphi \circ T_A^{\ell}$. For the transfer operators, it follows from (4.3) that

$$\widehat{T}^n \mathbb{1}_A = \sum_{k=1}^{n} \widehat{T}_A^k \mathbb{1}_{\{\varphi_k = n\}} \quad \text{and} \quad \sum_{k=1}^{n} \widehat{T}^k \mathbb{1}_A = \sum_{k=1}^{n} \widehat{T}_A^k \mathbb{1}_{\{\varphi_k \leq n\}}.$$

In particular, we have $a_n = \sum_{k=1}^{n} \mu|_A(\{\varphi_k \leq n\})$. Now, for the upper bound we have

$$\sum_{k=1}^{n} \widehat{T}^k \mathbb{1}_A = \sum_{k=1}^{n} \widehat{T}_A^k \mathbb{1}_{\{\varphi_k \leq n\}} \leq \sum_{k=1}^{n+m} \widehat{T}_A^k \mathbb{1}_{\{\varphi_k \leq n\}}$$

$$\leq m + \sum_{k=1}^{n} \widehat{T}_A^{k+m} \mathbb{1}_{\{\varphi_{k+m} \leq n\}} \leq m + \sum_{k=1}^{n} \widehat{T}_A^{k+m} \mathbb{1}_{\{\varphi_k \leq n\}}$$

$$\leq m + \sum_{k=1}^{n} (1 + \psi_m)\mu|_A(\{\varphi_k \leq n\}) = m + (1 + \psi_m)a_n,$$

where for the last inequality we used the fact that $\{\varphi_k \leq n\} \in \sigma(\mathbb{M}^k)$. Since this inequality holds for every $m, n \in \mathbb{N}$ and since (a_n) is diverging it follows that uniformly a.e.

$$\limsup_{n\to\infty} \frac{1}{a_n} \sum_{k=1}^{n} \widehat{T}^k \mathbb{1}_A \le 1.$$

For the lower bound we calculate similarly

$$\sum_{k=1}^{n} \widehat{T}^k \mathbb{1}_A \ge \sum_{k=1}^{n} \widehat{T}_A^{k+m} \mathbb{1}_{\{\varphi_{k+m} \le n\}} - m$$

$$\ge \sum_{k=1}^{n} \widehat{T}_A^{k+m} \mathbb{1}_{\{\varphi_k \le n\}} - \sum_{k=1}^{n} \widehat{T}_A^{k+m} \mathbb{1}_{\{\varphi_k \le n \le \varphi_{k+m}\}} - m$$

$$\ge \sum_{k=1}^{n} (1 - \psi_m) \mu|_A(\{\varphi_k \le n\}) - \sum_{k=1}^{n} \widehat{T}_A^{k+m} \mathbb{1}_{\{\varphi_k \le n \le \varphi_{k+m}\}} - m$$

$$\ge (1 - \psi_m) a_n - m - \sum_{k=1}^{n} \widehat{T}_A^{k+m} \mathbb{1}_{\{\varphi_k \le n \le \varphi_{k+m}\}}.$$

Now we observe that

$$\sum_{k=1}^{n} \widehat{T}_A^{k+m} \mathbb{1}_{\{\varphi_k \le n \le \varphi_{k+m}\}} \le (1 + \psi_0) \sum_{k=1}^{n} \mu|_A(\{\varphi_k \le n \le \varphi_{k+m}\})$$

$$= (1 + \psi_0) \sum_{k=1}^{n} \sum_{\ell=1}^{n} \mu|_A(\{\varphi_k = \ell;\ \varphi_m \circ T_A^k > n - \ell\})$$

$$\le (1 + \psi_0)^2 \sum_{k=1}^{n} \sum_{\ell=1}^{n} \mu|_A(\{\varphi_k = \ell\}) \mu|_A(\{\varphi_m > n - \ell\}).$$

Let us split up the last sum for $\ell = 1, \ldots, n - p$ and $\ell = n - p + 1, \ldots, n$ for a fixed $p < n$. For the first part we have

$$\sum_{k=1}^{n} \sum_{\ell=1}^{n-p} \mu|_A(\{\varphi_k = \ell\}) \mu|_A(\{\varphi_m > n - \ell\}) \le \mu|_A(\{\varphi_m > p\}) \sum_{k=1}^{n} \mu|_A(\{\varphi_k \le n - p\})$$

$$\le \mu|_A(\{\varphi_m > p\}) a_n.$$

For the second part, using $\mu|_A(\{\varphi_m > n - \ell\}) \le 1$ we have

$$\sum_{k=1}^{n} \sum_{\ell=n-p+1}^{n} \mu|_A(\{\varphi_k = \ell\}) \mu|_A(\{\varphi_m > n - \ell\})$$

$$\le \sum_{k=1}^{n} \mu|_A(\{n - p \le \varphi_k \le n\})$$

$$\leq \sum_{k=1}^{n} \mu|_A(\{\varphi_k \leq n\}) - \sum_{k=1}^{n} \mu|_A(\{\varphi_k \leq n-p\})$$

$$\leq \sum_{k=1}^{n} \mu|_A(\{\varphi_k \leq n\}) - \sum_{k=1}^{n-p} \mu|_A(\{\varphi_k \leq n-p\})$$

$$= a_n - a_{n-p} \leq p.$$

Combining these three inequalities we see that for all $m, p \in \mathbb{N}$ we have

$$\liminf_{n \to \infty} \frac{1}{a_n} \sum_{k=1}^{n} \widehat{T}^k \mathbb{1}_A \geq 1 - \psi_m - (1-\psi_0)^2 \mu|_A(\{\varphi_1 \geq p\}).$$

Since $\varphi_1 = \varphi$ is finite a.e. we have $\mu|_A(\{\varphi_1 \geq p\}) \to 0$ for $p \to \infty$. Also, $\psi_m \to 0$ for $m \to \infty$, this proves the uniform lower bound

$$\liminf_{n \to \infty} \frac{1}{a_n} \sum_{k=1}^{n} \widehat{T}^k \mathbb{1}_A \geq 1.$$

Combining these bounds finishes the proof. $\qquad\square$

Lemma 4.3.4. *For $A, B \in \mathcal{B}$ with $0 < \mu(A), \mu(B) < \infty$ and $A_n := A \setminus \bigcup_{k=1}^{n} T^{-k}(A)$ we have:*

(a) $\displaystyle\sum_{k=0}^{\infty} \widehat{T}^k \left(\mathbb{1}_{A_n}\right) = \mathbb{1}_X.$

(b) $\displaystyle\sum_{k=0}^{n} \widehat{T}^k \left(\mathbb{1}_B\right) = \sum_{\ell=0}^{n} \widehat{T}^\ell \left(\mathbb{1}_{A_\ell} \cdot \sum_{k=0}^{n-\ell} \widehat{T}^k \left(\mathbb{1}_B\right)\right) + \sum_{k=0}^{n} \widehat{T}^k \left(\mathbb{1}_{B \setminus \bigcup_{m=0}^{k} T^{-m}(A)}\right).$

Proof. To see the first claim, note that by Proposition 2.4.33 (a) we have for two measurable sets A, C with finite measure and $\mu(A) > 0$ that

$$\sum_{k=0}^{\infty} \mu \left(A_k \cap T^{-k}(C)\right) = \sum_{k=0}^{\infty} \mu \left(A \cap T^{-k}(C) \cap \{\varphi > k\}\right) = \mu(C).$$

Hence, using the Monotone Convergence Theorem, for every measurable set C with finite measure we have

$$\int_C \sum_{k=0}^{\infty} \widehat{T}^k(\mathbb{1}_{A_k}) d\mu = \sum_{k=0}^{\infty} \int \mathbb{1}_C \widehat{T}^k(\mathbb{1}_{A_k}) d\mu = \sum_{k=0}^{\infty} \mu \left(A_k \cap T^{-k}(C)\right) = \int_C \mathbb{1}_X d\mu.$$

This means that $\sum_{k=0}^{\infty} \widehat{T}^k(\mathbb{1}_{A_k}) = \mathbb{1}_X$.

The second claim follows in a similar way. For $n \in \mathbb{N}$ and two measurable sets A, C with finite measure we define

$$C_n := T^{-n}(C) \setminus \bigcup_{k=0}^{n} T^{-k}(A).$$

and observe that $T^{-1}(C_n) = C_{n+1} \cup (A \cap T^{-1}(C_n))$, where $C_{n+1} \cap (A \cap T^{-1}(C_n)) = \varnothing$. Now with A, B, C measurable sets with finite measure we have

$$\sum_{\ell=0}^{n} \sum_{k=0}^{n-\ell} \mu\left(B \cap T^{-k}(A_\ell) \cap T^{-(k+\ell)}(C)\right)$$

$$= \sum_{\ell=0}^{n} \sum_{k=0}^{n-\ell} \mu\left(B \cap T^{-k}\left(A \cap T^{-\ell}(C) \setminus \bigcup_{m=1}^{\ell} T^{-m}(A)\right)\right)$$

$$= \sum_{k=0}^{n} \mu\left(B \cap T^{-k}(C \cap A)\right) + \sum_{\ell=0}^{n} \sum_{k=0}^{n-\ell} \mu\left(B \cap T^{-k}\left(T^{-1}(C_{\ell-1}) \setminus B_\ell\right)\right).$$

Now for the second summand we find by a telescoping argument

$$\sum_{\ell=0}^{n} \sum_{k=0}^{n-\ell} \mu\left(B \cap T^{-k}\left(T^{-1}(C_{\ell-1}) \setminus B_\ell\right)\right)$$

$$= \sum_{\ell=0}^{n-1} \sum_{k=1}^{n-(\ell+1)+1} \mu(B \cap T^{-k}(C_\ell)) - \sum_{\ell=1}^{n} \sum_{k=0}^{n-\ell} \mu(B \cap T^{-k}(C_\ell))$$

$$= \sum_{\ell=0}^{n-1} \sum_{k=1}^{n-\ell} \mu(B \cap T^{-k}(C_\ell)) - \sum_{\ell=1}^{n-1} \sum_{k=1}^{n-\ell} \mu(B \cap T^{-k}(C_\ell)) - \sum_{\ell=1}^{n} \mu(B \cap C_\ell)$$

$$= \sum_{k=1}^{n} \mu(B \cap T^{-k}(C_0)) - \sum_{\ell=1}^{n} \mu(B \cap C_\ell)$$

$$= \sum_{k=0}^{n} \mu(B \cap T^{-k}(C \setminus A)) - \sum_{\ell=0}^{n} \mu(B \cap C_\ell)$$

Combining these two calculations gives

$$\sum_{k=0}^{n} \mu(B \cap T^{-k}(C)) = \sum_{\ell=0}^{n} \sum_{k=0}^{n-\ell} \mu\left(B \cap T^{-k}(A_\ell) \cap T^{-(k+\ell)}(C)\right)$$

$$+ \sum_{\ell=0}^{n} \mu\left(B \cap T^{-\ell} C \setminus \bigcup_{m=0}^{\ell} T^{-m} A\right).$$

The remaining part of the proof is analogous to the first part. □

Proposition 4.3.5. *Let $(X, \mathcal{B}, \mu, T)$ be a conservative ergodic measure-preserving system, and let $A, B \in \mathcal{B}$ with $0 < \mu(A), \mu(B) < \infty$. Suppose that there exists an increasing positive sequence $(a_n)_{n \geq 1}$ tending to infinity such that a.e. on A we have*

$$\lim_{n \to \infty} \frac{1}{a_n} \sum_{k=0}^{n} \widehat{T}^k \mathbb{1}_B = \mu(B). \tag{4.4}$$

Then the system is pointwise dual ergodic. In particular, the existence of a Darling–Kac set implies pointwise dual ergodicity.

Proof. We are going to prove that the convergence in (4.4) in fact holds a.e. on X. Then Hurewicz's Ergodic Theorem combined with this observation proves pointwise dual ergodicity.

First note that by the Chacon–Ornstein Lemma 4.1.4 and the assumption stated in the proposition we have for all $N \in \mathbb{N}$ that

$$\frac{a_{n-N}}{a_n} = \frac{a_{n-N}}{a_n} \frac{\sum_{k=0}^{n} \widehat{T}^k \mathbb{1}_B}{\sum_{k=0}^{n-N} \widehat{T}^k \mathbb{1}_B} \left(1 - \frac{\sum_{k=n-N+1}^{n} \widehat{T}^k \mathbb{1}_B}{\sum_{k=0}^{n} \widehat{T}^k \mathbb{1}_B}\right) \to 1.$$

As before, for $n \geq 0$, let $A_n := A \setminus \bigcup_{k=1}^{n} T^{-k}A$. By Egorov's Theorem we may assume without loss of generality that the convergence in (4.4) holds uniformly on A. Fix $\varepsilon > 0$. Using the first part of Lemma 4.3.4 we find for almost every $x \in X$ an $n_0 \in \mathbb{N}$ such that $\sum_{k=0}^{n_0} \widehat{T}^k \mathbb{1}_{A_k}(x) \geq (1 - \varepsilon)$. Further there exists $n_1 > n_0$ such that for all $n \geq n_1$ and uniformly on A we have

$$\frac{1}{a_{n-n_0}} \sum_{k=0}^{n-n_0} \widehat{T}^k(\mathbb{1}_B) \geq (1 - \varepsilon)\mu(B) \quad \text{and} \quad \frac{a_{n-n_0}}{a_n} \geq (1 - \varepsilon).$$

Now using the second part of Lemma 4.3.4 gives for all $n \geq n_1$

$$\frac{1}{a_n} \sum_{k=0}^{n} \widehat{T}^k \mathbb{1}_B(x) = \frac{1}{a_n} \left(\sum_{k=0}^{n} \widehat{T}^k \left(\mathbb{1}_{A_k} \cdot \sum_{\ell=0}^{n-k} \widehat{T}^\ell \mathbb{1}_B\right) + \sum_{k=0}^{n} \widehat{T}^k \mathbb{1}_{B \setminus \bigcup_{j=0}^{k} T^{-j}(A)}\right)(x)$$

$$\geq \frac{1}{a_n} \sum_{k=0}^{n_0} \widehat{T}^k \left(\mathbb{1}_{A_k} \cdot \sum_{\ell=0}^{n-k} \widehat{T}^\ell(\mathbb{1}_B)\right)(x)$$

$$\geq \frac{1}{a_n} \sum_{k=0}^{n_0} \widehat{T}^k \left(\mathbb{1}_{A_k} \cdot \sum_{\ell=0}^{n-n_0} \widehat{T}^\ell(\mathbb{1}_B)\right)(x)$$

$$\geq (1 - \varepsilon)\frac{a_{n-n_0}}{a_n}\mu(B) \sum_{k=0}^{n_0} \widehat{T}^k \left(\mathbb{1}_{A_k}\right)(x) \geq (1 - \varepsilon)^3\mu(B).$$

This shows that a.e. we have $\liminf_{n \to \infty} \frac{1}{a_n} \sum_{k=0}^{n} \widehat{T}^k \mathbb{1}_B \geq \mu(B)$.

Towards the upper bound for the limit superior, fix $\varepsilon > 0$. By Hurewicz's Ergodic Theorem we also have a.e. on A,

$$\lim_{n \to \infty} \frac{1}{a_n} \sum_{k=0}^{n} \widehat{T}^k \mathbb{1}_A = \mu(A). \tag{4.5}$$

Again by Egorov's Theorem we find a set $A' \in \mathcal{B}$ such that $\mu(A') > (1 + \varepsilon)^{-1}\mu(A)$ and the convergence in (4.5) holds uniformly on A'. Hence there exists $n_0 \in \mathbb{N}$ such that on A' and all $n \geq n_0$ we have $\frac{1}{a_n} \sum_{k=0}^{n} \widehat{T}^k \mathbb{1}_A \leq (1 + \varepsilon)\mu(A)$ for all $n \geq n_0$. Using the second part of Lemma 4.3.4 again with A' in the place of B gives

$$
\frac{1}{a_n}\sum_{k=0}^{n}\widehat{T}^k\mathbb{1}_{A'} = \frac{1}{a_n}\sum_{k=0}^{n}\widehat{T}^k\left(\mathbb{1}_{A'_k}\cdot\sum_{\ell=0}^{n-k}\widehat{T}^\ell\mathbb{1}_{A'}\right)
$$

$$
\leq \sum_{k=0}^{n}\widehat{T}^k\left(\mathbb{1}_{A'_k}\cdot\frac{1}{a_n}\sum_{\ell=0}^{n}\widehat{T}^\ell(\mathbb{1}_A)\right)
$$

$$
\leq (1+\varepsilon)\mu(A)\sum_{k=0}^{n}\widehat{T}^k\mathbb{1}_{A'_k}
$$

$$
\leq (1+\varepsilon)\mu(A) \leq (1+\varepsilon)^2\mu(A').
$$

Hence, we have a.e. $\limsup_{n\to\infty}\frac{1}{a_n}\sum_{k=0}^{n}\widehat{T}^k\mathbb{1}_{A'} \leq \mu(A')$. Another application of Hurewicz's Ergodic Theorem allows us to replace $\mathbb{1}_{A'}$ by $\mathbb{1}_B$ in the above inequality. This finishes the proof. $\qquad\square$

4.4 Exercises

Exercise 4.4.1. Let $X := (0,\infty)$ and λ denote the Lebesgue measure restricted to X. Consider

$$
V_1: L_1(\lambda)\to L_1(\lambda)
$$
$$
f\mapsto \left(x\mapsto V_1(f)(x) := e^{-x}f(x)\right).
$$

Is V_1 a positive contractive operator? Is this operator conservative? Determine the sets $\{S_\infty g = \infty\}$ and $\{S_\infty g > 0\}$ for some integrable $g > 0$. Is V_1 ergodic?

Exercise 4.4.2. Let $X := (0,\infty)$ and λ denote the Lebesgue measure restricted to X. Consider

$$
V_2: L_1(\lambda)\to L_1(\lambda)
$$
$$
f\mapsto \left(x\mapsto V_2(f)(x) := \mathbb{1}_{(0,1)}\int f\,d\lambda\right).
$$

Is V_2 a positive contractive operator? Is this operator conservative? Determine the sets $\{S_\infty g = \infty\}$ and $\{S_\infty g > 0\}$ for some integrable $g > 0$. Is V_2 ergodic?

Exercise 4.4.3. Let $f, g \in L_1(\mu)$ with $g \geq 0$. Prove with the help of Wiener's Maximal Inequality that we have

$$
\lim_{n\to\infty}\frac{S_n f}{S_n g} = \frac{S_\infty f}{S_\infty g}.
$$

exists μ-a.e. on $\{S_\infty g > 0\} \setminus \{S_\infty g = \infty\}$. Go back to Exercises 4.4.1 and 4.4.2 and consider the limit $\lim_{n\to\infty} S_n f / S_n g$ on $\{S_\infty g > 0\}$ and $\{S_\infty g = \infty\}$, respectively.

Exercise 4.4.4. Let V be a positive conservative contraction on $L_1(\mu)$. Then a measurable set A is called V^*-*invariant* if $\mathbb{1}_A$ is a V^*-invariant function. Do the V^*-invariant sets form a σ-algebra? Compare this collection with the σ-algebra generated by the V^*-invariant functions.

Exercise 4.4.5. For $g \in L_1^+(\mu)$ we have that the set $\{S_\infty g = \infty\}$ is V^*-invariant.

Exercise 4.4.6. Prove Remark 4.1.15.

Exercise 4.4.7. Let $(X, \mathcal{B}, \mu)$ be a probability space. We recall that a sub-σ-algebra $\mathcal{F} \subset \mathcal{B}$ is said to be trivial if for all $A \in \mathcal{F}$ we have $\mu(F) \in \{0, 1\}$. Show that for all $f \in L_1(\mu)$ we have

$$\mathbb{E}_\mu(f|\mathcal{F}) = \int f \, d\mu.$$

Exercise 4.4.8. Prove the statement in (4.2).

5 Applications of infinite ergodic theory

In this chapter, we shall first consider the sum-level sets for the continued fraction expansion and prove all the corresponding results to those obtained in Chapter 3 for the α-sum-level sets. In the continued fraction case investigated here, though, we will first have to prove that the Gauss map has the ψ-mixing property, in order to apply the results from infinite ergodic theory given in Chapter 4, as the renewal arguments given in Chapter 3 are no longer sufficient.

We will also see an application of the fine asymptotics of the Lebesgue measure of the sum-level sets to Diophantine approximation. Finally, we employ infinite ergodic theory to show that the even Stern–Brocot sequence is uniformly distributed with respect to certain canonical weightings.

5.1 Sum-level sets for the continued fraction expansion, first investigations

To begin, let us recall the definition of the sum-level sets $(C_n)_{n\geq 1}$ for the continued fraction expansion:

$$C_n := \left\{ [x_1, x_2, x_3, \ldots] \in [0, 1] : \sum_{i=1}^{k} x_i = n \text{ for some } k \in \mathbb{N} \right\}$$

$$= \bigcup_{k=1}^{n} \bigcup_{(x_1,\ldots,x_k):\sum_{i=1}^{k} x_i = n} C(x_1, \ldots, x_k).$$

We claimed in the introduction to this chapter that the renewal theory arguments as used in Chapter 3 are no longer sufficient to analyse the sum-level sets for the continued fraction expansion. To see why, observe that if we wanted to prove a result equivalent to Lemma 3.2.8 in this situation, we would be doomed to failure, as the following calculation shows:

$$\frac{3}{10} = \lambda(C_3) \neq \frac{1}{2} \cdot \lambda(C_2) + \frac{1}{6} \cdot \lambda(C_1) + \frac{1}{12} \cdot \lambda(C_0) = \frac{1}{3}.$$

(Recall that we calculated the values of the Lebesgue measure of the first four sum-level sets at the beginning of Chapter 3.)

Before stating the first main theorem, we need the following lemma which provides the crucial link between the sequence of sum-level sets and the Farey map. Note that this lemma contradicts our initial impression that the sequence of sum-level sets is not a dynamical entity, despite its apparent strangeness.

Lemma 5.1.1. *For all $n \in \mathbb{N}$, we have that*

$$F^{-(n-1)}(\mathcal{C}_1) = \mathcal{C}_n.$$

Proof. By computing the images of $\mathcal{C}_1$ under the inverse images F_0 and F_1 of the Farey map, one immediately verifies that $F^{-1}(\mathcal{C}_1) = \mathcal{C}_2$. We then proceed by way of induction as follows. Assume that for some $n \in \mathbb{N}$ we have that $F^{-(n-1)}(\mathcal{C}_1) = \mathcal{C}_n$. Since $F^{-n}(\mathcal{C}_1) = F^{-1}(F^{-(n-1)}(\mathcal{C}_1)) = F^{-1}(\mathcal{C}_n)$, it is then sufficient to show that $F^{-1}(\mathcal{C}_n) = \mathcal{C}_{n+1}$. For this, let $x = [x_1, x_2, x_3 \ldots] \in \mathcal{C}_n$ be given. Then there exists $\ell \in \mathbb{N}$ such that

$$x \in C(x_1, \ldots, x_\ell) \quad \text{and} \quad \sum_{i=1}^{\ell} x_i = n.$$

By computing the images of x under the inverse branches F_0 and F_1, one obtains that $F^{-1}(x) = \{[1, x_1, x_2, \ldots], [x_1 + 1, x_2, \ldots]\}$. Since we have that

$$1 + \sum_{i=1}^{\ell} x_i = (x_1 + 1) + \sum_{i=2}^{\ell} x_i = n + 1,$$

this shows that $F^{-1}(x) \subset \mathcal{C}_{n+1}$, and hence, $F^{-1}(\mathcal{C}_n) \subset \mathcal{C}_{n+1}$. The reverse inclusion $\mathcal{C}_{n+1} \subset F^{-1}(\mathcal{C}_n)$ can be established by counting the Stern–Brocot intervals contained in $\mathcal{C}_{n+1}$. $\square$

Remark 5.1.2. Notice that an analogous result also holds for the α-sum-level sets, but that this observation was not necessary for the analysis given in Section 3.2.2.

We are now almost ready to prove our first main theorem. The proof will follow on combining Lemma 5.1.1 with the next result which depends on the fact that F is exact (as shown in Theorem 2.5.6), so that we can apply Lin's criterion for exactness (see Theorem 2.5.7). Let us also recall that the unique absolutely continuous invariant measure for the Farey map, denoted by ν_F, is given by $\nu_F(A) := \int_A h_F(x)\, d\lambda(x)$, where $h_F(x) := 1/x$ (see Proposition 2.3.19).

Proposition 5.1.3. *For each measurable set C which satisfies $\nu_F(C) < \infty$, we have that*

$$\lim_{n \to \infty} \lambda\left(F^{-n}(C)\right) = 0.$$

Proof. Let $C \in \mathcal{B}$ be given as stated in the proposition. So, for each $A \in \mathcal{B}$ for which $0 < \nu_F(A) < \infty$, we then have

$$\lambda\left(F^{-n}(C)\right) = \nu_F\left(\mathbb{1}_{F^{-n}(C)} \cdot h_F^{-1}\right) = \nu_F\left(\mathbb{1}_C \circ F^n \cdot h_F^{-1}\right)$$

$$= \nu_F\left(\mathbb{1}_C \circ F^n \cdot \left(h_F^{-1} - \frac{\mathbb{1}_A}{\nu_F(A)} + \frac{\mathbb{1}_A}{\nu_F(A)}\right)\right)$$

$$\leq \left\| \widehat{F}^n \left(h_F^{-1} - \frac{\mathbb{1}_A}{\nu_F(A)} \right) \right\|_1 + \frac{\nu_F \left(F^{-n}(C) \cap A \right)}{\nu_F(A)}$$

$$\leq \left\| \widehat{F}^n \left(h_F^{-1} - \frac{\mathbb{1}_A}{\nu_F(A)} \right) \right\|_1 + \frac{\nu_F(C)}{\nu_F(A)}$$

$$\to \frac{\nu_F(C)}{\nu_F(A)}, \text{ for } n \text{ tending to infinity.}$$

Here, the limit follows from the fact that $\nu_F \left(\left(h_F^{-1} - \mathbb{1}_A / \nu_F(A) \right) \right) = 0$ and F is exact, and hence, Lin's criterion is applicable. Therefore, by choosing $A \in \mathcal{B}$ such that $\nu_F(A)$ is arbitrarily large, the proposition follows. $\qquad\square$

We can now easily apply this result to determine the limit of the sequence $(\lambda(\mathcal{C}_n))_{n\geq 1}$, which was our first objective.

Theorem 5.1.4.

$$\lim_{n\to\infty} \lambda(\mathcal{C}_n) = 0.$$

Proof. The proof follows immediately by first putting $C = \mathcal{C}_1$ in Proposition 5.1.3, and then using the fact that $\mathcal{C}_n = F^{-(n-1)}(\mathcal{C}_1)$, for all $n \in \mathbb{N}$, as shown in Lemma 5.1.1. $\qquad\square$

Remark 5.1.5.
1. The above theorem (and proof) can be found in [KS12b]. Also in that paper, there is another proof of the same theorem, which is more elementary, in the sense that it uses less infinite ergodic theory. The other proof depends upon first showing that $\liminf_{n\to\infty} \lambda(\mathcal{C}_n) = 0$. This fact was first established by Fiala and Kleban [FK10], but they did not provide a proof for the limit.
2. The arguments given above can be slightly modified to give a different proof of the infinite-type part of Theorem 3.2.9. Note that this will not work for the finite measure case.

5.2 ψ-mixing for the Gauss map and the Gauss problem

Recall that in Corollary 2.5.9, we used Lin's criterion for exactness to show that if a map T is exact, then it is also mixing (cf. Definition 2.5.8). It follows, in light of Theorem 2.4.12, that the Gauss map G is mixing. Here, we want to show that G satisfies the stronger property of ψ-mixing which was introduced in Chapter 4 (see Definition 4.3.2). This property can sometimes, for instance in [Aar97], be found under the name *continued-fraction mixing* precisely because the Gauss map satisfies it. The proof that we will present here is very much inspired by the proof given in [Ios92].

Before proving that the Gauss map G is ψ-mixing, we must make some preliminary remarks. First of all, recall that we have $\widehat{G}\mathbb{1} = \mathbb{1}$. Secondly, we remind the reader that the identity $\widehat{G}f = h_G^{-1} P_G (h_G f)$ was established in Proposition 2.3.18, where there we understand the operator P_G as acting on $L^1(\mu)$. Throughout this section we want to use the pointwise definition of $h_G^{-1} P_G (h_G f)$ for a concrete measurable function f (and in a later section we will do similarly for the Farey map F), but in order to shorten the notation, we will simply write $\widehat{G}(f)(x) = h_G^{-1} P_G (h_G f)(x)$. Thus, for the operator $\widehat{G}$ for all $x \in [0, 1]$ and "suitable" functions f, where the class of suitable functions will be defined below, we will write

$$\widehat{G}f(x) = \sum_{i=1}^{\infty} p_i(x) f\left(\frac{1}{i+x}\right),$$

where we have set $p_i(x) := (1 + x)/((i + x)(i + 1 + x))$, for all $i \in \mathbb{N}$. (With the pointwise definition above kept in mind, the proof of this fact is simply a calculation; we leave it to Exercise 5.7.2.) Notice that $(p_i(x))_{i \geq 1}$ is a probability vector; this will turn out to be crucial later.

Further, let us introduce the set of functions of *bounded variation*,

$$\mathrm{BV} := \{f : [0, 1] \to \mathbb{R} : \mathrm{var} f < \infty\},$$

where $\mathrm{var} f$ is defined to be $\mathrm{var} f := \mathrm{var}_{[0,1]} f$ and where $\mathrm{var}_{[a,b]} f$ is defined, for $[a, b] \subset [0, 1]$, to be

$$\mathrm{var}_{[a,b]} f := \sup\left\{\sum_{i=1}^{n} |f(x_{i+1}) - f(x_i)| : a \leq x_1 < \cdots < x_{n+1} \leq b, n \in \mathbb{N}\right\}.$$

Note that any function of bounded variation is in particular bounded. One can show that any function of bounded variation f can be written as the difference $g - h$ of two bounded functions g and h, which are either both non-decreasing or both non-increasing (you are asked to prove this in Exercise 5.7.1). More precisely, these functions can be chosen, for $x \in [0, 1]$, in the non-decreasing case to be

$$g(x) := \mathrm{var}_{[0,x]} f \quad \text{and} \quad h(x) := \mathrm{var}_{[0,x]} f - f(x),$$

and in the non-increasing case to be

$$g(x) := \mathrm{var}_{[x,1]} f \quad \text{and} \quad h(x) := \mathrm{var}_{[x,1]} f - f(x).$$

It is easy to see, and we leave it as an exercise, that in the non-decreasing case, $\mathrm{var} g = g(1) - g(0) = \mathrm{var} f$ and $\mathrm{var} h = h(1) - h(0) = \mathrm{var} f + f(0) - f(1)$, and in the non-increasing case, $\mathrm{var} g = \mathrm{var} f$ and $\mathrm{var} h = \mathrm{var} f - f(0) + f(1)$.

Lemma 5.2.1. *If $f : [0, 1] \to \mathbb{R}$ is bounded and either non-decreasing or non-increasing, then $\widehat{G}f$ is bounded and non-increasing or non-decreasing, respectively.*

Proof. That $\widehat{G}f$ is bounded follows directly from the fact that $\sum_{i=1}^{\infty} p_i(x) = 1$ for all $x \in [0, 1]$. For the proof of the remaining assertions, assume that f is non-decreasing (for the non-increasing case consider $-f$ instead) and let $x, y \in [0, 1]$ be fixed such that $x < y$. Then,

$$\widehat{G}f(x) - \widehat{G}f(y) = \sum_{i=1}^{\infty} p_i(x)f\left(\frac{1}{x+i}\right) - \sum_{i=1}^{\infty} p_i(y)f\left(\frac{1}{y+i}\right)$$

$$= \sum_{i=1}^{\infty} (p_i(x) - p_i(y))f\left(\frac{1}{x+i}\right)$$

$$+ \sum_{i=1}^{\infty} p_i(y)\left(f\left(\frac{1}{x+i}\right) - f\left(\frac{1}{y+i}\right)\right)$$

$$\geq \sum_{i=1}^{\infty} (p_i(x) - p_i(y))f\left(\frac{1}{x+i}\right)$$

$$= \sum_{i=1}^{\infty} (p_i(x) - p_i(y))\left(f\left(\frac{1}{x+i}\right) - f\left(\frac{1}{x+2}\right)\right) \geq 0,$$

where, in the final equality, and inequality, we have used the observations that $\sum_{i=1}^{\infty}(p_i(x) - p_i(y)) = 0$, that p_1 is decreasing, and that p_i is increasing for all $i \geq 3$. □

Lemma 5.2.2. *For each monotone and bounded function $f : [0, 1] \to \mathbb{R}$, we have that*

$$\mathrm{var}\,\widehat{G}f \leq \frac{1}{2}\,\mathrm{var}f.$$

Proof. Using Lemma 5.2.1 and assuming first that f is non-decreasing, we obtain

$$\mathrm{var}\,\widehat{G}f := \widehat{G}f(0) - \widehat{G}f(1) = \sum_{i=1}^{\infty} p_i(0)f\left(\frac{1}{i}\right) - \sum_{i=1}^{\infty} p_i(1)f\left(\frac{1}{1+i}\right)$$

$$= \frac{1}{2}f(1) - \sum_{i=2}^{\infty}(p_{i-1}(1) - p_i(0))f\left(\frac{1}{i}\right)$$

$$= \frac{1}{2}f(1) - \frac{1}{2}\sum_{i=2}^{\infty}\frac{2}{i(i+1)}f\left(\frac{1}{i}\right) \leq \frac{1}{2}f(1) - \frac{1}{2}f(0) = \frac{1}{2}\mathrm{var}f,$$

where we used the facts that f is non-decreasing and that $(2/(i(i+1)))_{i\geq 2}$ is a probability vector. For the non-increasing case, consider $-f$ instead and observe that $\mathrm{var}(-f) = \mathrm{var}f$ and $\widehat{G}(-f) = -\widehat{G}f$. □

Remark 5.2.3. Note that the constant $1/2$ in the latter lemma is optimal. In order to see this, choose f to be any non-decreasing function such that $f|_{[0,1/2]} = 0$ and $0 < f(1) < \infty$. For this choice, the above calculation immediately shows that $\mathrm{var}\,\widehat{G}f = 1/2\,\mathrm{var}f$.

Now we are in a position to take the next step towards the proof that G is ψ-mixing, namely, we will obtain a bound on the distance, arising from the supremum norm

$\| \cdot \|_\infty$, of the n-th iterate of $\widehat{G}$ applied to functions of bounded variation and the integral of these functions.

Lemma 5.2.4. *For each $f \in \mathrm{BV}$ and $n \in \mathbb{N}$, we have that*

$$\left\| \widehat{G}^n f - \int f \, dm_G \right\|_\infty \le 2^{-n} (2 \operatorname{var} f - |f(0) - f(1)|).$$

If f is monotone then we have $2 \operatorname{var} f - |f(0) - f(1)| = \operatorname{var} f$.

Proof. First observe that for each $x \in [0, 1]$ and $f \in \mathrm{BV}$, we have that

$$\left| f(u) \right| - \left| \int f \, dm_G \right| \le \left| \int f(u) - f(x) \, dm_G(x) \right| \le \operatorname{var} f,$$

and hence, $\|f\|_\infty \le \left| \int f \, dm_G \right| + \operatorname{var} f$. It follows from (2.4) by setting $f := 1$ that $\int \widehat{G}^n f \, dm_G = \int f \, dm_G$, and thus the above observation gives, for each $n \in \mathbb{N}$,

$$\left\| \widehat{G}^n f - \int f \, dm_G \right\|_\infty \le \operatorname{var} \left(\widehat{G}^n f - \int f \, dm_G \right) = \operatorname{var} \widehat{G}^n f.$$

Let $f = g - h$, for the monotone bounded functions g and h as defined prior to Lemma 5.2.1. Using Lemma 5.2.2, it follows that

$$\left\| \widehat{G}^n f - \int f \, dm_G \right\|_\infty \le \left\| \widehat{G}^n g - \int g \, dm_G \right\|_\infty + \left\| \widehat{G}^n h - \int h \, dm_G \right\|_\infty$$

$$\le \operatorname{var} \widehat{G}^n g + \operatorname{var} \widehat{G}^n h$$

$$\le 2^{-n} (\operatorname{var} g + \operatorname{var} h)$$

$$= 2^{-n} \min \{2 \operatorname{var} f - f(0) + f(1), 2 \operatorname{var} f + f(0) - f(1)\}$$

$$= 2^{-n} (2 \operatorname{var} f - |f(0) - f(1)|) \qquad \square$$

Lemma 5.2.5. *For each $B \in \mathcal{B}$, all $m, n \in \mathbb{N}_0$ and either every Gauss cylinder set $C :=$ $C(x_1, \ldots, x_n)$ of level $n > 0$ or for $n = 0$ and $C := [0, 1]$, we have that*

$$\left| \lambda \left(G^{-n-m}(B) \cap C \right) - m_G(B) \lambda(C) \right| \le 2^{-m} \log 2 \, m_G(B) \lambda(C).$$

Proof. Using Lemma 5.2.4 and, as before, the relationship between $\widehat{G}$ and P_G, we have that

$$\left| \lambda \left(G^{-n-m}(B) \cap C \right) - m_G(B) \lambda(C) \right|$$

$$= \left| \int \left((\mathbb{1}_B \circ G^{m+n}) \cdot h_G^{-1} \cdot \mathbb{1}_C \right) dm_G - m_G(B) \lambda(C) \right|$$

$$= \left| \int \left(\mathbb{1}_B \cdot \widehat{G}^{m+n} \left(h_G^{-1} \mathbb{1}_C \right) - \lambda(C) \mathbb{1}_B \right) dm_G \right|$$

$$\le \int \mathbb{1}_B \, dm_G \left\| \widehat{G}^m \widehat{G}^n \left(h_G^{-1} \mathbb{1}_C \right) - \lambda(C) \right\|_\infty$$

$$\leq m_G(B) 2^{-m} \left(2\operatorname{var}\left(\widehat{G}^n\left(h_G^{-1}\mathbb{1}_C\right)\right) - \left|\widehat{G}^n\left(h_G^{-1}\mathbb{1}_C\right)(0) - \widehat{G}^n\left(h_G^{-1}\mathbb{1}_C\right)(1)\right|\right)$$

$$= m_G(B) 2^{-m} \left(2\operatorname{var}\left(h_G^{-1}P_G^n(\mathbb{1}_C)\right) - \left|\left(h_G^{-1}P_G^n(\mathbb{1}_C)\right)(0) - \left(h_G^{-1}P_G^n(\mathbb{1}_C)\right)(1)\right|\right)$$

$$\leq 2^{-m}\log 2\, m_G(B)\lambda(C).$$

Here, the final inequality can be seen as follows: Directly from the definition of P_G, we have that $P_G^n(\mathbb{1}_C)$ is equal to the derivative of the inverse branch of G^n that maps the unit interval onto the Gauss cylinder C. With $p_n/q_n := [x_1, \dots, x_n]$, this derivative is given for $y \in [0, 1]$ by $1/(q_n + yq_{n-1})^2$ (the proof of this fact is left to Exercise 5.7.3). Thus we obtain the formula

$$\left(h_G^{-1}P_G^n(\mathbb{1}_C)\right)(y) = \frac{\log 2\,(1+y)}{(q_n + yq_{n-1})^2}.$$

To shorten the notation, let us set $f_n := \left(h_G^{-1}P_G^n(\mathbb{1}_C)\right)$. To obtain an upper bound on the variation of this function, we first take the derivative:

$$f_n'(y) = \frac{1}{(q_n + yq_{n-1})^2}\left(1 - \frac{2q_{n-1}(1+y)}{q_n + yq_{n-1}}\right),$$

and then observe that if $x_n = 1$, this derivative is always negative (so the function is monotonically decreasing), if $x_n \geq 3$, then the derivative is always positive (so the function is monotonically increasing), and if $x_n = 2$ we have that the function has a maximum at the point $y = q_n/q_{n-1} - 2 \in (0, 1)$. Therefore, in the cases $x_n = 1$ and $x_n \geq 3$, we have immediately that

$$\operatorname{var}(f_n) = \left|f_n(0) - f_n(1)\right| = \left|\frac{\log 2}{q_n^2} - \frac{2\log 2}{(q_n + q_{n-1})^2}\right|$$

$$\leq \frac{\log 2}{q_n^2 + q_nq_{n-1}}\left|\frac{(q_{n-1}/q_n)^2 + 2q_{n-1}/q_n - 1}{1 + q_{n-1}/q_n}\right|$$

$$\leq \log 2\,\lambda(C)\max\left\{\left|\frac{x^2 + 2x - 1}{1 + x}\right| : x \in [0, 1]\right\}$$

$$= \log 2\,\lambda(C).$$

If $x_n = 2$, or, equivalently, if $q_n/q_{n-1} \in [2, 3]$, a similar calculation shows that

$$\operatorname{var}(f_n) \leq 2f_n(q_n/q_{n-1} - 2) - f_n(0) - f_n(1)$$

$$\leq \log 2\,\lambda(C)\max\left\{\left|\frac{x^4 - 4x^3 + 3x^2 + 2x + 2}{2x(x+1)(x-1)}\right| : x \in [2, 3]\right\}$$

$$= \frac{1}{6}\log 2\,\lambda(C).$$

The case in which $C := [0, 1]$ and $n = 0$ is more straightforward and is left to Exercise 5.7.4. $\qquad\square$

Corollary 5.2.6. *For every set $B \in \mathcal{B}$, we have that*

$$\lim_{n \to \infty} \lambda(G^{-n}(B)) = m_G(B).$$

Proof. This is an immediate consequence of Lemma 5.2.5 with $C = [0, 1]$. □

Theorem 5.2.7. *The system $([0, 1], \mathcal{B}, m_G, G)$ is ψ-mixing with respect to the partition α_H. More precisely, for all positive integers $m, n \in \mathbb{N}$, any Gauss cylinder set $C :=$ $C(x_1, \ldots, x_n)$ of level n and every set $B \in \mathcal{B}$, we have that*

$$\left| m_G \left(G^{-n-m} B \cap C \right) - m_G(B) m_G(C) \right| \leq 2^{-m} \log 2 \, m_G(B) m_G(C).$$

Proof. Let B and C be given as stated in the theorem. For $y := (y_1, \ldots, y_N) \in \mathbb{N}^N$, we use the notation $[yC]$ to denote the Gauss cylinder set $C(y_1, \ldots, y_N, x_1, \ldots, x_n)$ of level $n + N$. With this notation, Lemma 5.2.5 implies that for each $N \in \mathbb{N}$, we have

$$\left| \lambda \left(G^{-N} \left(G^{-n-m} B \cap C \right) \right) - m_G(B) \lambda \left(G^{-N} C \right) \right|$$

$$= \left| \sum_{y \in \mathbb{N}^N} \left(\lambda \left([yC] \cap G^{-n-m-N} B \right) - m_G(B) \lambda([yC]) \right) \right|$$

$$\leq 2^{-m} \log 2 \, m_G(B) \sum_{y \in \mathbb{N}^N} \lambda([yC])$$

$$= 2^{-m} \log 2 \, m_G(B) \lambda \left(G^{-N}(C) \right).$$

Letting N tend to infinity in the above inequality and invoking Corollary 5.2.6 finishes the proof of the theorem. □

Remark 5.2.8. The history of ψ-mixing for the Gauss map is a rather long one, and proving ψ-mixing for the Gauss map can be considered to be the first problem in the metric theory of continued fractions. It originates in a letter which Gauss wrote to Laplace on the 30th of January 1812, asking him to give an estimate of the error term

$$\rho_n(x) := |\lambda(G^{-n}([0, x])) - m_G([0, x])|, \text{ for } x \in [0, 1] \text{ and } n \in \mathbb{N}.$$

This problem, known as the *generalised Gauss problem*[1], remained open for a long time, until R. O. Kuzmin [Kuz28] and P. Lévy [Lev29] independently and almost simultaneously gave a solution. Kuzmin showed that $\rho_n \leq c\kappa^{\sqrt{n}}$, for some positive constants $\kappa < 1$ and c, whereas Lévy obtained the result that $\rho_n \leq c\theta^n$, for positive constants $\theta < 0.68 \ldots$ and c. These results of Kuzmin and Lévy were then followed by various improvements by several authors, among them W. Doeblin, F. Schweiger and P. Szüsz. The currently most satisfying estimate has been obtained

---

1 Note that the actual *Gauss problem* was to show what we have obtained in Corollary 5.2.6.

by E. Wirsing [Wir74], who showed that the constant θ in Lévy's estimate is equal to 0.30366300289873265860....

5.3 Pointwise dual ergodicity for the Farey map

First let us recall from Chapter 2 the induced map F_{C_1} of the Farey map on the interval $C_1 := [1/2, 1]$, which was defined in Example 2.4.30. This map acts on points $[1, x_2, x_3, \ldots]$ in C_1 by

$$F_{C_1}([1, x_2, x_3, \ldots]) = [1, x_3, x_4, \ldots].$$

We showed in the proof of Proposition 2.4.31 that this induced system is measure-theoretically isomorphic to the Gauss system. Given that G is ψ-mixing, as was shown in the previous section (see Theorem 5.2.7), it follows immediately that the map F_{C_1} is also ψ-mixing. Before stating the first result, recall that pointwise dual ergodicity and Darling–Kac sets were introduced in Definitions 4.2.1 and 4.3.1, respectively.

Lemma 5.3.1. *The set C_1 is a Darling–Kac set for the Farey map.*

Proof. This follows directly from Proposition 4.3.3, in combination with the discussion above. □

Theorem 5.3.2. *The system $([0, 1], \mathcal{B}, \nu_F, F)$ is pointwise dual ergodic.*

Proof. Since, according to Lemma 5.3.1, the map F has a Darling–Kac set, the result follows directly from Theorem 4.3.5. □

Now we know that F is pointwise dual ergodic, we can obtain information about wandering rates and return sequences. Let us denote the return sequence associated to F by $(v_n)_{n\geq1}$. In order to determine the asymptotic type of (v_n), we must compute the wandering rate $(w_n(C_1))_{n\geq1}$, (see Definition 4.2.3), where we recall that the wandering rate $(w_n(C))$ for any $C \in \mathcal{B}$ is given by

$$w_n(C) := \nu_F \left(\bigcup_{k=1}^{n} F^{-(k-1)}(C) \right).$$

So, for each $n \in \mathbb{N}$ we have

$$w_n(C_1) = \nu_F \left(\bigcup_{k=1}^{n} T^{-(k-1)}(C_1) \right) = \nu_F \left(\mathbb{1}_{[1/(n+1),1]} \right) = \log(n+1) \sim \log(n).$$

Next, observe that this wandering rate is slowly varying at infinity, where we recall from Chapter 3 that this means

$$\lim_{n\to\infty} w_{k\cdot n}(C_1)/w_n(C_1) = 1, \text{ for each } k \in \mathbb{N}.$$

Lemma 5.3.3. *For the map F, the return sequence (v_n) can be defined by setting*

$$v_n := \frac{n}{\log(n)}.$$

Proof. This follows on combining Proposition 4.2.8 with the fact that F is pointwise dual ergodic. □

5.4 Uniform and uniformly returning sets

In this section, it will turn out to be helpful to have a formula for the transfer operator $\widehat{F} := \widehat{F}_{\nu_F}$, as we had for $\widehat{G}$ in Section 5.2. Exactly as before, we have that for $f \in L_1(\nu_F)$,

$$\widehat{F}(f) = h_F^{-1} P_F(h_F f).$$

It is then straightforward to check that this leads to the pointwise definition

$$\widehat{F}(f)(x) = \frac{1}{1+x} f\left(\frac{x}{1+x}\right) + \frac{x}{1+x} f\left(\frac{1}{1+x}\right).$$

Now, we aim to obtain uniform sets for the system $([0,1], \mathcal{B}, \nu_F, F)$. First recall the general definition of a uniform set stated in Definition 4.2.4: Let $(X, \mathcal{B}, \mu, T)$ be pointwise dual ergodic with return sequence (r_n). Then, a set $A \in \mathcal{B}$ with positive, finite measure is called a *uniform set* for $f \in L_1^+(\mu)$ if

$$\frac{1}{r_n} \sum_{k=0}^{n-1} \widehat{T}^k f \to \int f \, d\mu \text{ uniformly } (\bmod \, \mu) \text{ on } A.$$

We introduce a set of functions $\mathcal{D}$, in order to show that the set $\mathcal{C}_1$ is uniform for all of the elements of $\mathcal{D}$. So, define

$$\mathcal{D} := \{f \in L_1^+(\nu_F) \cap C^2([0,1]) : f' > 0 \text{ and } f'' \le 0\}.$$

We remark that $\mathcal{D}$ is not empty, since $\mathrm{id}_{[0,1]} \in \mathcal{D}$.

Lemma 5.4.1. *We have $\widehat{F}(\mathcal{D}) \subset \mathcal{D}$.*

Proof. Let $f \in \mathcal{D}$ and consider the derivative of $\widehat{F}(f)$: By the monotonicity of f and f' we have

$$\widehat{F}(f)'(x) = \underbrace{\frac{f'\left(\frac{x}{x+1}\right) - xf'\left(\frac{1}{x+1}\right)}{(x+1)^3}}_{>0} + \underbrace{\frac{f\left(\frac{1}{x+1}\right) - f\left(\frac{x}{x+1}\right)}{(x+1)^2}}_{>0},$$

which implies $\widehat{F}(f)' > 0$. Furthermore, an easy calculation shows

$$\widehat{F}(f)''(x) = \frac{f''\left(\frac{x}{x+1}\right) + xf''\left(\frac{1}{x+1}\right)}{(x+1)^5} + \frac{2\left(f\left(\frac{x}{x+1}\right) - f\left(\frac{1}{x+1}\right)\right)}{(x+1)^3}$$

$$+ \frac{2(x+1)\left((x-1)f'\left(\frac{1}{x+1}\right) - 2f'\left(\frac{x}{x+1}\right)\right)}{(x+1)^5} \le 0.$$

This finishes the proof. □

Lemma 5.4.2. *The set C_1 is uniform for every $f \in \mathcal{D}$.*

Proof. Fix $f \in \mathcal{D}$. Note that $\widehat{F}f(1) = f(1/2)$. Lemma 5.4.1 implies that $x \mapsto \widehat{F}^n f(x)$ is monotone increasing. Thus, recalling from Lemma 5.3.3 that $v_n = n/\log(n)$, we have for every $x \in C_1$

$$\frac{1}{v_n}\sum_{k=0}^n \widehat{F}^k f(1/2) \le \frac{1}{v_n}\sum_{k=0}^n \widehat{F}^k f(x) \le \frac{1}{v_n}\sum_{k=0}^n \widehat{F}^k f(1)$$

$$\le \frac{1}{v_n}\sum_{k=0}^{n+1} \widehat{F}^k f(1) \le \frac{1}{v_n}f(1) + \frac{1}{v_n}\sum_{k=0}^n \widehat{F}^k f(1/2)$$

Since $\lim_{n\to\infty} v_n^{-1}f(1) = 0$ the uniform convergence $v_n^{-1}\sum_{k=0}^n \widehat{F}^k f \to \int f \, d\nu_F$ follows from pointwise dual ergodicity. □

We will now introduce a somewhat stronger notion than uniform sets, namely, that of uniformly returning sets.

Definition 5.4.3. Let $(X, \mathcal{B}, \mu, T)$ be a conservative, ergodic, measure-preserving system. A set $C \in \mathcal{B}$ with $0 < \mu(C) < \infty$ is called *uniformly returning* for $f \in L_1^+(\mu)$ if there exists an increasing sequence $(w_n) := (w_n(f, C))_{n \ge 1}$ of positive real numbers such that μ-almost everywhere and uniformly on C we have that

$$\lim_{n\to\infty} w_n \widehat{T}^n(f) = \int f \, d\mu.$$

Remark 5.4.4. If A, B are sets of finite μ-measure and B is uniformly returning for $\mathbb{1}_A$, we immediately obtain a form of mixing for infinite systems, in the sense that

$$w_n\mu(A \cap T^{-n}(B)) = \int w_n \widehat{T}^n(\mathbb{1}_A) \cdot \mathbb{1}_B \, d\mu \to \mu(A)\mu(B).$$

Note that this could be considered to be a little unsatisfactory as a definition, since it is basically considering an infinite system by restricting it to a set of finite measure. The work of Lenci, mentioned directly before Definition 2.5.8, addresses this issue by looking at the problem of defining infinite mixing on the entire space from a more physical point of view.

Let us now return to the Farey map.

Proposition 5.4.5. *For the Farey system $([0,1], \mathcal{B}, \nu_F, F)$ we have that if $v \in L_1(\nu_F)$ satisfies*

$$\lim_{n \to \infty} w_n \widehat{F}^n(v) = \int v \, d\nu_F \text{ almost everywhere uniformly on } \mathcal{C}_1,$$

then the same holds on any compact subset of $(0, 1]$.

Proof. Let us first recall that, for $x \in (0, 1]$ and $n \in \mathbb{N}$,

$$\left(P_F^{n+1}(h_F \cdot v) \right)(x) = P_F \left(P_F^n(h_F \cdot v) \right)(x)$$

$$= \left(P_F^n(h_F \cdot v) \right)(F_0(x)) \cdot |F_0'(x)|$$

$$+ \left(P_F^n(h_F \cdot v) \right)(F_1(x)) \cdot |F_1'(x)|,$$

which gives

$$\left(P_F^n(h_F \cdot v) \right)(F_0(x)) = \frac{(P_F^{n+1}(h_F \cdot v))(x) - (P_F^n(h_F \cdot v))(F_1(x)) \cdot |F_1'(x)|}{|F_0'(x)|}. \tag{5.1}$$

We proceed by induction as follows. The start of the induction is given by the assumption in the theorem. For the inductive step, assume that the statement holds for $\bigcup_{i=1}^k \mathcal{C}_i$, for some $k \in \mathbb{N}$. Then consider some arbitrary $y \in \mathcal{C}_{k+1}$, and let x denote

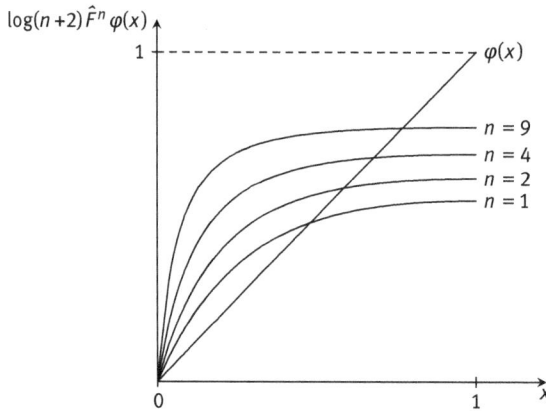

Fig. 5.1. The iterates of the Farey transfer operator $\widehat{F}^n$ acting on $\varphi : x \mapsto x$ and rescaled with the wandering rate approximate a.s. uniformly on compact subsets of $(0, 1]$ the constant function of height $1 = \int \varphi \, d\mu$.

the unique element in A_k such that $F_0(x) = y$. Using (5.1), the fact that $\widehat{F} = h_F^{-1}P_F(h_F \cdot v)$ and the inductive hypothesis in tandem with the assumption that $\lim w_n/w_{n+1} = 1$, we obtain that

$$w_n\left(\widehat{F}^n(v)\right)(y) = w_n\left(\widehat{F}^n(v)\right)(F_0(x)) = \frac{w_n(P_F^n(h_F \cdot v))(F_0(x))}{h_F(F_0(x))}$$

$$= \frac{w_n(P_F^{n+1}(h_F \cdot v))(x) - |F_1'(x)| \cdot w_n(P_F^n(h_a \cdot v))(F_1(x))}{h_F(F_0(x)) \cdot |F_0'(x)|}$$

$$\sim \frac{h_F(x) - h_F(F_1(x)) \cdot |F_1'(x)|}{h_a(F_0(x)) \cdot |F_0'(x)|} \int v d\nu_F = \int v d\nu_F,$$

where the last equality is a consequence of the eigenequation $P_F h_F = h_F$. □

Let us now return to our collection $\mathcal{D}$. To prove that $\mathcal{C}_1$ is indeed uniformly returning for every element of $\mathcal{D}$ we need the following observation.

Lemma 5.4.6. On $\mathcal{C}_1$, for every $f \in \mathcal{D}$ we have that

$$\widehat{F}^n f < \widehat{F}^{n-1} f, \quad \text{for all } n \in \mathbb{N}.$$

Proof. Fix $f \in \mathcal{D}$. Again we use the fact that $\widehat{F}f(1) = f(1/2)$. Then for every $x \in \mathcal{C}_1$ we have

$$\widehat{F}^n f(x) \le \max\left\{\widehat{F}^n f(x) : x \in \mathcal{C}_1\right\} = \widehat{F}^n f(1) = \widehat{F}^{n-1} f(1/2)$$

$$= \min\left\{\widehat{F}^{n-1} f(x) : x \in \mathcal{C}_1\right\} \le \widehat{F}^{n-1} f(x),$$

where at least one of the inequalities must be strict. □

Proposition 5.4.7. The set $\mathcal{C}_1$ is uniformly returning for every $f \in \mathcal{D}$. That is for all $f \in \mathcal{D}$ we have

$$\log(n)\widehat{F}^n(f) \to \int f d\nu_F, \quad \text{uniformly on } \mathcal{C}_1.$$

(See Fig. 5.1 for an illustration).

Proof. Let $\lambda, \eta \in \mathbb{R}$ be arbitrary fixed real numbers with $0 < \lambda < \eta < \infty$. Putting

$$V_n := \sum_{k=0}^n \widehat{F}^k(f),$$

we have by the monotonicity of the sequence $\left(\widehat{F}^n(f)|_{\mathcal{C}_1}\right)_{n \in \mathbb{N}}$ that

$$\frac{\widehat{F}^{\lfloor n\eta \rfloor}(f)}{V_n} \cdot (\lfloor n\eta \rfloor - \lfloor n\lambda \rfloor) \le \frac{V_{\lfloor n\eta \rfloor} - V_{\lfloor n\lambda \rfloor}}{V_n} \le \frac{\widehat{F}^{\lfloor n\lambda \rfloor}(f)}{V_n} \cdot (\lfloor n\eta \rfloor - \lfloor n\lambda \rfloor).$$

Since $\lfloor n\eta \rfloor - \lfloor n\lambda \rfloor \sim n(\eta - \lambda)$ as $n \to \infty$, we have for fixed $\varepsilon \in (0,1)$ and all n sufficiently large

$$n(1-\varepsilon)(\eta - \lambda) \leq \lfloor n\eta \rfloor - \lfloor n\lambda \rfloor \leq n(1+\varepsilon)(\eta - \lambda).$$

This implies for all n sufficiently large

$$\frac{n\widehat{F}^{\lfloor n\eta \rfloor}(f)}{V_n} \cdot (1-\varepsilon)(\eta - \lambda) \leq \frac{V_{\lfloor n\eta \rfloor} - V_{\lfloor n\lambda \rfloor}}{V_n} \leq \frac{n\widehat{F}^{\lfloor n\lambda \rfloor}(f)}{V_n} \cdot (1+\varepsilon)(\eta - \lambda).$$

Since

$$\frac{V_{\lfloor n\eta \rfloor} - V_{\lfloor n\lambda \rfloor}}{V_n} \to \eta^{\alpha} - \lambda^{\alpha} \quad \text{as } n \to \infty \quad \mu\text{-a.e.} \quad \text{uniformly on } \mathcal{C}_1,$$

we obtain on the one hand

$$\frac{1}{1+\varepsilon} \cdot \frac{\eta^{\alpha} - \lambda^{\alpha}}{\eta - \lambda} \leq \liminf_{n\to\infty} \frac{n\widehat{F}^{\lfloor n\lambda \rfloor}(f)}{V_n} \quad \mu\text{-a.e.} \quad \text{uniformly on } \mathcal{C}_1.$$

Letting $\eta \to \lambda$ and $\varepsilon \to 0$, it follows that

$$\alpha\lambda^{\alpha-1} \leq \liminf_{n\to\infty} \frac{n\widehat{F}^{\lfloor n\lambda \rfloor}(f)}{V_n} \quad \mu\text{-a.e.} \quad \text{uniformly on } \mathcal{C}_1.$$

On the other hand, we obtain similarly

$$\limsup_{n\to\infty} \frac{n\widehat{F}^{\lfloor n\eta \rfloor}(f)}{V_n} \leq \alpha\eta^{\alpha-1} \quad \mu\text{-a.e.} \quad \text{uniformly on } \mathcal{C}_1.$$

Since λ and η are arbitrary, we have for any $c > 0$

$$\frac{n\widehat{F}^{\lfloor nc \rfloor}(f)}{V_n} \to \alpha c^{\alpha-1} \quad \mu\text{-a.e.} \quad \text{uniformly on } \mathcal{C}_1.$$

Finally using $V_{\lfloor nc \rfloor} \sim c^{\alpha} V_n$ μ-a.e. uniformly on $\mathcal{C}_1$ and $\lfloor nc \rfloor \sim cn$, we obtain for $m = \lfloor nc \rfloor$

$$\frac{m\widehat{F}^m(f)}{V_m} = \frac{\lfloor nc \rfloor}{n} \cdot \frac{n\widehat{F}^{\lfloor nc \rfloor}(f)}{V_n} \cdot \frac{V_n}{V_{\lfloor nc \rfloor}} \to \alpha \quad \mu\text{-a.e.} \quad \text{uniformly on } \mathcal{C}_1.$$

From this and Lemma 5.4.2 the assertion follows. $\qquad\square$

Remark 5.4.8. The material developed here for the set of functions $\mathcal{D}$ can be found in greater generality in the context of operator renewal theory in various works, including [Gou11, Gou04, MT15, MT12, Sar02, KKSS15, KKS15].

5.5 Finer asymptotics of Lebesgue measure of sum-level sets

Using the results obtained in the previous sections, we are now in a position to state and prove our first result on the finer asymptotics of the sum-level sets.

Theorem 5.5.1.

$$\sum_{k=1}^{n} \lambda(\mathcal{C}_k) \sim \frac{n}{\log_2 n}.$$

Proof. First, recall from Lemma 5.1.1 that $\mathcal{C}_k = F^{-(k-1)}(\mathcal{C}_1)$. Therefore,

$$\frac{1}{v_n} \cdot \sum_{k=1}^{n} \lambda(\mathcal{C}_k) = \frac{1}{v_n} \int \sum_{k=0}^{n-1} \mathbb{1}_{\mathcal{C}_1} \circ F^k \, d\lambda$$

$$= \frac{1}{v_n} \int \sum_{k=0}^{n-1} \mathbb{1}_{\mathcal{C}_1} \circ F^k \cdot h_F^{-1} \, d\nu_F$$

$$= \int \mathbb{1}_{\mathcal{C}_1} \frac{1}{v_n} \left(\sum_{k=0}^{n-1} \widehat{F}^k(h_F^{-1}) \right) d\nu_F.$$

Since $\mathcal{C}_1$ is a uniform set for $h_F^{-1} \in \mathcal{D}$ we have that $\lim_{n \to \infty} v_n^{-1} \sum_{k=0}^{n-1} \left(\widehat{F}^k(h_F^{-1}) \right) = \int h_F^{-1} d\nu_F = 1$ a.e. uniformly on $\mathcal{C}_1$, and so

$$\lim_{n \to \infty} \frac{1}{v_n} \cdot \sum_{k=1}^{n} \lambda(\mathcal{C}_k) = \log 2.$$

$\square$

Our second, and final, theorem concerning the Lebesgue measure of the sum-level sets gives a significant improvement of Theorem 5.1.4 and Theorem 5.5.1. That is, by increasing the dosage of infinite ergodic theory, we are able to obtain the following sharp estimate for the asymptotic behaviour of the Lebesgue measure of the sum-level sets.

Theorem 5.5.2.

$$\lambda(\mathcal{C}_n) \sim \frac{\log 2}{\log n}.$$

Proof. We use again the fact from Lemma 5.1.1 that $\mathcal{C}_k = F^{-(k-1)}(\mathcal{C}_1)$. Therefore,

$$w_n \cdot \lambda(\mathcal{C}_n) = w_n \int \mathbb{1}_{\mathcal{C}_1} \circ F^n \, d\lambda$$

$$= w_n \int \mathbb{1}_{\mathcal{C}_1} \circ F^n \cdot h_F^{-1} \, d\nu_F$$

$$= \int \mathbb{1}_{\mathcal{C}_1} \left(w_n \widehat{F}^n(h_F^{-1}) \right) d\nu_F.$$

Since C_1 is uniformly returning for h_F^{-1} we have that

$$\lim_{n \to \infty} w_n \left(\widehat{F}^n(h_F^{-1}) \right) = \int h_F^{-1} \mathrm{d}\nu_F = 1$$

a. e. uniformly on C_1, and so

$$\lim_{n \to \infty} w_n \cdot \lambda(C_n) = \log 2.$$

□

Remark 5.5.3. We refer the interested reader to [Hee15] for an effective bound on the error term of this asymptotic.

Let us now give an application of the above theorem to elementary metrical Diophantine analysis. We first state the following result, which is a consequence of Theorem 1.2.19, given in Section 1.2.4.

Theorem 5.5.4. *For λ-almost every $x = [x_1, x_2, x_3, \ldots] \in [0, 1]$, we have that*

$$\limsup_{n \to \infty} \frac{\log(x_n/n)}{\log \log n} = 1.$$

Proof. In light of Corollary 1.2.20, for λ-a.e. $x = [x_1, x_2, x_3, \ldots] \in [0, 1]$ we have that for all $\varepsilon > 0$ and all sufficiently large $n \in \mathbb{N}$,

$$x_n < n(\log n)^{1+\varepsilon}.$$

Taking logarithms of each side, it follows that for all sufficiently large $n \in \mathbb{N}$,

$$\log(x_n) < \log(n) + (1 + \varepsilon) \log \log(n)$$

or, rewriting this expression, again for all sufficiently large $n \in \mathbb{N}$, we have

$$\frac{\log(x_n/n)}{\log \log(n)} < 1 + \varepsilon.$$

Hence,

$$\limsup_{n \to \infty} \frac{\log(x_n/n)}{\log \log(n)} \leq 1 + \varepsilon.$$

Since $\varepsilon > 0$ was arbitrary, we may conclude that

$$\limsup_{n \to \infty} \frac{\log(x_n/n)}{\log \log(n)} \leq 1.$$

On the other hand, it also follows from Corollary 1.2.20 that for Lebesgue-almost every $x = [x_1, x_2, x_3, \ldots] \in [0, 1]$, we have that for infinitely many $n \in \mathbb{N}$,

$$x_n > n \log(n).$$

Then,

$$1 \le \limsup_{n\to\infty} \frac{\log(x_n/n)}{\log\log(n)}$$

and the theorem is proved. □

In contrast to the almost-everywhere property of continued fraction digits stated in Theorem 5.5.4, we can now prove a similar statement associated to the Farey coding using Theorem 5.5.2. Note that $\sum_{i=1}^{n} x_i$ represents the word length associated with the Farey system, whereas the parameter n represents the word length associated with the Gauss system. Before stating the theorem, let us remind the reader that we use the notation $A \asymp B$ to mean that there exists a constant $c \ge 1$ such that $c^{-1}A \le B \le cA$.

Proposition 5.5.5. *For λ-almost every $x = [x_1, x_2, x_3, \ldots] \in [0, 1]$, we have that*

$$\limsup_{n\to\infty} \frac{\log(x_{n+1}/\sum_{i=1}^{n} x_i)}{\log\log(\sum_{i=1}^{n} x_i)} \le 0.$$

Proof. For each $n \in \mathbb{N}$ and $\varepsilon > 0$, let

$$A_n^{\varepsilon} := \bigcup_{k\in\mathbb{N}} \left\{ C(x_1, \ldots, x_{k+1}) : \sum_{i=1}^{k} x_i = n, x_{k+1} \ge n(\log n)^{\varepsilon} \right\}$$

and define

$$\mathcal{A}_n^{\varepsilon} := \bigcup_{C\in A_n^{\varepsilon}} C.$$

Now recall from (1.8) that

$$\lambda(C(x_1, \ldots, x_k, m)) \asymp m^{-2}\lambda(C(x_1, \ldots, x_k)).$$

Therefore, for all $k, \ell \in \mathbb{N}$, we obtain that

$$\sum_{x_{k+1}\ge\ell} \lambda(C(x_1, \ldots, x_k, x_{k+1})) \asymp \ell^{-1}\lambda(C(x_1, \ldots, x_k)).$$

Using this estimate and Theorem 5.5.2, we deduce that

$$\lambda(\mathcal{A}_n^{\varepsilon}) = \sum_{k=1}^{n} \sum_{\substack{(x_1,\ldots,x_k) \\ \sum_{i=1}^{k} x_i=n}} \sum_{x_{k+1}\ge n(\log n)^{\varepsilon}} \lambda(C(x_1, \ldots, x_k, x_{k+1}))$$

$$\asymp \sum_{k=1}^{n} \sum_{\substack{(x_1,\ldots,x_k) \\ \sum_{i=1}^{k} x_i=n}} \frac{\lambda(C(x_1, \ldots, x_k))}{n(\log n)^{\varepsilon}}$$

$$= \frac{1}{n(\log n)^\varepsilon} \sum_{k=1}^{n} \sum_{\substack{(x_1,\ldots,x_k) \\ \sum_{l=1}^{k} x_l = n}} \lambda(C(x_1,\ldots,x_k))$$

$$= \frac{\lambda(C_n)}{n(\log n)^\varepsilon} \sim \frac{\log 2}{n(\log n)^{1+\varepsilon}}.$$

Hence, since the above calculation implies that the series $\sum_{n=1}^{\infty} \lambda(\mathcal{A}_n^\varepsilon)$ converges, a straightforward application of the Borel–Cantelli Lemma then yields, where $\mathcal{A}_\infty^\varepsilon :=$ $\limsup_{n\to\infty} \mathcal{A}_n^\varepsilon$, that

$$\lambda(\mathcal{A}_\infty^\varepsilon) = 0, \text{ for each } \varepsilon > 0.$$

On considering the complement of the set $\mathcal{A}_\infty^\varepsilon$ in $[0,1]$, we have now shown that, for each $\varepsilon > 0$ and for λ-almost all $x = [x_1, x_2, x_3, \ldots]$,

$$x_{k+1} < \left(\sum_{i=1}^{k} x_i\right) \left(\log \sum_{i=1}^{k} x_i\right)^\varepsilon, \text{ for all } k \in \mathbb{N} \text{ sufficiently large.}$$

By taking logarithms on both sides of the above inequality, we obtain, for all sufficiently large $k \in \mathbb{N}$, that

$$\frac{\log(x_{k+1}) - \log\left(\sum_{i=1}^{k} x_i\right)}{\log\log\left(\sum_{i=1}^{k} x_i\right)} = \frac{\log\left(x_{k+1}/\sum_{i=1}^{k} x_i\right)}{\log\log\left(\sum_{i=1}^{k} x_i\right)} < \varepsilon.$$

It therefore follows that

$$\limsup_{k\to\infty} \frac{\log\left(x_{k+1}/\sum_{i=1}^{k} x_i\right)}{\log\log\left(\sum_{i=1}^{k} x_i\right)} \leq \varepsilon.$$

Finally, on letting ε tend to zero, the lemma follows. □

Remark 5.5.6. Using these ideas, there are other results related to continued fractions and Diophantine analysis that can be obtained. For instance, for the random variable

$$X_n(x) := \max\left\{\sum_{i=1}^{k} x_i : \sum_{i=1}^{k} x_i \leq n, \ k \in \mathbb{N}_0\right\}, \quad x \in \mathbb{I},$$

the process $n - X_n$ is investigated in [KS08a]. In that paper, a uniform law and large deviation law are derived.

For further interesting results in the context of continued fraction digit sums we also refer to [GLJ93, GLJ96], wherein alternating sums of continued fraction digits are considered.

5.6 Uniform distribution of the even Stern–Brocot sequence

Let us begin this section by recalling the concept of *weak convergence* of probability measures. We say a sequence $(\mu_n)_{n\in\mathbb{N}}$ of Borel probability measures on $(\mathbb{R}, \mathcal{B})$ converges weakly to a Borel probability measure μ if for all $f \in C_b(\mathbb{R})$ we have

$$\lim_{n\to\infty} \int f \, d\mu_n = \int f \, d\mu.$$

For this we write w-$\lim_n \mu_n = \mu$. There are different effective ways to check weak convergence despite the difficulty of trying to directly use the definition. We will need the following two characterisations, which can be found in the standard literature on probability, for example in [Gut13]. We recall that $\Delta_\mu : x \mapsto \mu((-\infty, x])$ denotes the distribution function of μ.

- (Distribution function) w-$\lim_n \mu_n = \mu$ if and only if $\lim_{n\to\infty} \Delta_{\mu_n}(x) = \Delta_\mu$ for every $x \in \mathbb{R}$ that is a continuity point of Δ_μ.
- (Method of Moments) For a probability measure μ, the function

$$t \mapsto \int \exp(t \cdot x) \, d\mu$$

is called the *moment generating function* (you will see why in Exercise 5.7.6). We have that if $\int \exp(t \cdot x) \, d\mu_n \to \int \exp(t \cdot x) \, d\mu \in \mathbb{R}$ for n tending to infinity and for all t in a neighbourhood of 0, then w-$\lim_n \mu_n = \mu$.

Let δ_x denote the Dirac measure in x, that is $\delta_x(A) = 1$ for $x \in A$ and $\delta_x(A) = 0$ otherwise. Our aim in this section is to prove the following theorem:

Theorem 5.6.1. *For each rational number $v/w \in (0, 1]$ we have that*

$$\text{w-}\lim_{n\to\infty} \ \log(n^{vw}) \sum_{p/q\in F^{-n}\{v/w\}} q^{-2} \delta_{p/q} = \lambda. \tag{5.2}$$

In order to prove this result, we first prove the following proposition, and then one further lemma, which will allow us to transfer the result stated below for intervals to the atomic measures considered in Theorem 5.6.1.

Proposition 5.6.2. *For each interval $[a, b] \subset (0, 1]$ we have that*

$$\text{w-}\lim_{n\to\infty} \left(\frac{\log n}{\log (b/a)} \cdot \lambda|_{F^{-n}([a,b])} \right) = \lambda.$$

Proof. Consider the family of functions $(\varphi_t)_{t\in[-1,1]}$ given by $\varphi_t : x \mapsto x \cdot \exp(t \cdot x)$. The first aim is to show that for all $t \in [-1, 1]$ we have

$$\widehat{F}\varphi_t \in \mathcal{D}.$$

Indeed, for $t \in [-1, 0]$ this is an immediate consequence of $\widehat{F}(\mathcal{D}) \subset \mathcal{D}$ (see Lemma 5.4.1) by noting that φ_t is increasing, concave with $\varphi_t(0) = 0$, that is $\varphi_t \in \mathcal{D}$. For $t \in (0, 1]$, a straightforward computation shows that the first derivative of $\widehat{F}\varphi_t$ at $x \in [0, 1]$ is given by

$$\left(\widehat{F}\varphi_t\right)'(x) = \frac{\varphi_t'\left(\frac{x}{x+1}\right) - x\varphi_t'\left(\frac{1}{x+1}\right)}{(x+1)^3} + \frac{\varphi_t\left(\frac{1}{x+1}\right) - \varphi_t\left(\frac{x}{x+1}\right)}{(x+1)^2}.$$

For the second derivative we then obtain

$$\left(\widehat{F}\varphi_t\right)''(x) = \frac{\left(-2xt - 6x + 2t + xt^2 + 2x^3 - 4tx^2 - 4\right)\exp\left(\frac{tx}{x+1}\right)}{(x+1)^6}$$

$$+ \frac{\left(2tx - 6x - 2t + xt^2 + 2x^3 + 4tx^2 - 4\right)\exp\left(\frac{t}{x+1}\right)}{(x+1)^6}.$$

This immediately implies that $\left(\widehat{F}\varphi_t\right)'' \leq 0$, for all $t \in (0, 1]$. Hence, $\widehat{F}\varphi_t$ is concave and we have that $\left(\widehat{F}\varphi_t\right)'$ is decreasing on $[0, 1]$. Since $\left(\widehat{F}\varphi_t\right)'(1) = 0$, this shows that on $[0, 1]$ we have that $\left(\widehat{F}\varphi_t\right)' \geq 0$. Hence, $\widehat{F}\varphi_t \in \mathcal{D}$, for all $t \in [-1, 1]$.

We proceed by noting that Proposition 5.4.7 combined with Proposition 5.4.5 guarantees that every compact interval contained in $(0, 1]$ is a uniformly returning set for φ_t, for each $t \in [-1, 1]$. In order to complete the proof of the proposition, we employ the method of moments as follows. For each $[a, b] \subset (0, 1]$ and for each $t \in [-1, 1]$, we have

$$\lim_{n\to\infty} \int \exp(tx) \cdot \frac{\log n}{\nu_F([a,b])} \cdot \mathbb{1}_{F^{-n}([a,b])}(x)\, d\lambda(x)$$

$$= \lim_{n\to\infty} \frac{\log n}{\nu_F([a,b])} \cdot \nu_F\left(\varphi_t \cdot \mathbb{1}_{F^{-n}([a,b])}\right) = \lim_{n\to\infty} \frac{\log n}{\nu_F([a,b])} \cdot \nu_F\left(\widehat{F}^n \varphi_t \cdot \mathbb{1}_{[a,b]}\right)$$

$$= \nu_F(\varphi_t) = \int \exp(tx)\, d\lambda(x).$$

This shows that, restricted to $[-1, 1]$, the moment generating functions for the sequence of measures $\left(\frac{\log n}{\log(b/a)} \cdot \lambda|_{F^{-n}([a,b])}\right)_{n\in\mathbb{N}}$ converge to the moment generating function for the Lebesgue measure λ. In turn, this shows the weak convergence in question and hence finishes the proof of Proposition 5.6.2. $\qquad\square$

For the next lemma we introduce the notation Φ for the free semi-group generated by the inverse branches F_0 and F_1 of the Farey map F. Note that for each rational number $v/w \in (0, 1]$ we have that

$$\{F^{-n}\{v/w\} : n \in \mathbb{N}\} = \{g(v/w) : g \in \Phi\}.$$

Moreover, note that the Φ-orbit of 1 is equal to the set of rational numbers contained in $(0, 1)$. (Note that this is just a slightly different way of repeating what we already knew from Section 1.3, when obtaining the Farey coding of the rational numbers.) Then if we associate matrices to the inverse branches $F_0 : x \mapsto \frac{x}{1+x}$ and $F_1 : x \mapsto \frac{1}{1+x}$, and observe that

$$\left\langle \begin{pmatrix} 1 & 0 \\ 1 & 1 \end{pmatrix}, \begin{pmatrix} 0 & 1 \\ 1 & 1 \end{pmatrix} \right\rangle \subset GL_2(\mathbb{Z}) := \left\{ \begin{pmatrix} a & b \\ c & d \end{pmatrix} : a, b, c, d \in \mathbb{Z}, |ad - bc| = 1 \right\},$$

then to each $g \in \Phi$ we can associate a matrix $\begin{pmatrix} a & b \\ c & d \end{pmatrix}$ from $GL_2(\mathbb{Z})$. The action of g on $\mathbb{C}$ is given by $g : z \mapsto \frac{az+b}{cz+d}$. Thus $g(1) = v/w$, for some $v, w \in \mathbb{N}$ such that $v < w$ and $\gcd(v, w) = 1$. Furthermore, for the modulus of the derivative of g at x we have that $|g'(x)| = |cx + d|^2$. To make this more precise, see also Exercise 5.7.5.

In the following we let $\mathcal{U}_\varepsilon(x)$ denote the interval centred at $x \in \mathbb{R}$ of Euclidean diameter $\mathrm{diam}(\mathcal{U}_\varepsilon(x))$ equal to $\varepsilon > 0$.

Lemma 5.6.3. *For each $g \in \Phi$ there exists a constant C_g such that for all $\varepsilon > 0$ sufficiently small and for all $h \in \Phi$, we have*

$$\left| \mathrm{diam}(h(\mathcal{U}_\varepsilon(g(1)))) - \varepsilon \, |(h'(g(1))| \right| \le \varepsilon C_g \, \mathrm{diam}(h(\mathcal{U}_\varepsilon(g(1)))).$$

Proof. First we will prove the following bounded distortion property. Using Exercise 5.7.5, we know that for each $g \in \Phi$ we can find $m, n \in \mathbb{N}$ such that $|g'(z)| = |mz + n|^2$. Now fix $z \in (0, 1)$ and let $0 < \varepsilon < z/2$ and $x, y \in \mathcal{U}_\varepsilon(z)$. Then

$$\sup_{g \in \Phi} \left| \frac{|g'(x)|}{|g'(y)|} - 1 \right| \le \sup_{m,n \in \mathbb{N}} \left| \frac{(my + n)^2}{(mx + n)^2} - 1 \right| \le \sup_{m,n \in \mathbb{N}} \left| \frac{m^2(y^2 - x^2) + 2nm(y - x)}{(mx + n)^2} \right|$$

$$\le \sup_{m,n \in \mathbb{N}} \frac{(2m^2 + 2nm)}{(mz/2 + n)^2} |y - x| \le \frac{8}{z^2} |x - y| \sup_{m,n \in \mathbb{N}} \frac{m^2 + mn}{(m + 2n)^2}$$

$$\le \frac{16}{z^2} |x - y|.$$

Here, the last inequality can be seen by treating the two cases $m \le n$ and $m > n$ separately. Now, fix $g \in \Phi$. Then we have, for $0 < \varepsilon < g(1)^2/32$ and each $h \in \Phi$,

$$\left| \frac{\varepsilon|h'(g(1))|}{\mathrm{diam}(h(\mathcal{U}_\varepsilon(g(1))))} - 1 \right| \le \left| \frac{1}{\frac{1}{\varepsilon} \int_{g(1)-\varepsilon/2}^{g(1)+\varepsilon/2} \frac{|h'(\eta)|}{|h'(g(1))|} - 1\, d\eta + 1} - 1 \right|$$

$$\le \left| \frac{1}{1 - 16\varepsilon/g(1)^2} - 1 \right| \le \frac{32\varepsilon}{g(1)^2}.$$

From this we deduce that

$$\left| \mathrm{diam}(h(\mathcal{U}_\varepsilon(g(1)))) - \varepsilon|h'(g(1))| \right| < \mathrm{diam}(h(\mathcal{U}_\varepsilon(g(1)))) \frac{32\varepsilon}{g(1)^2}$$

Setting $C_g := 32/g(1)^2$ then finishes the proof. $\qquad\square$

Proof of Theorem 5.6.1. Let $g \in \Phi$ be given and define, for $\varepsilon > 0$ sufficiently small,

$$\mathcal{U}_{g,\varepsilon,n} := F^{-(n-1)}\left(\mathcal{U}_\varepsilon(g(1))\right).$$

With $u_{g,\varepsilon} := 1/\nu_F(\mathcal{U}_\varepsilon(g(1))) = 1/\log\left((g(1)+\varepsilon/2)/(g(1)-\varepsilon/2)\right)$, consider the scaled and restricted Lebesgue measure $\nu_{g,\varepsilon,n}$ which is given, for each $n \in \mathbb{N}$, by

$$\nu_{g,\varepsilon,n} := u_{g,\varepsilon} \log n \cdot \lambda|_{\mathcal{U}_{g,\varepsilon,n}}.$$

By Proposition 5.6.2, we then have that w-$\lim_{n\to\infty} \nu_{g,\varepsilon,n} = \lambda$. Then observe that

$$\lim_{\varepsilon \searrow 0} \varepsilon u_{g,\varepsilon} = \lim_{\varepsilon \searrow 0} \frac{\varepsilon}{\log \dfrac{g(1)+\varepsilon/2}{g(1)-\varepsilon/2}} = g(1), \qquad (5.3)$$

and consider the measures $\rho_{g,n}$ defined, for each $n \in \mathbb{N}$, by

$$\rho_{g,n} := g(1)\log n \sum_{f(1)\in F^{-(n-1)}\{g(1)\}} \frac{|f'(1)|}{|g'(1)|} \cdot \delta_{f(1)}.$$

Using Lemma 5.6.3, we now obtain the following for all $x \in [0,1]$, where $\Delta_{g,\varepsilon,n}^{(\nu)}$ and $\Delta_{g,n}^{(\rho)}$ denote the distribution functions of the measures $\nu_{g,\varepsilon,n}$, and $\rho_{g,n}$, respectively.

$$\left| \Delta_{g,\varepsilon,n}^{(\nu)}(x) - \Delta_{g,n}^{(\rho)}(x) \right|$$

$$\le \frac{1}{\varepsilon} \log n \sum_{\substack{h\in\Phi:\\ hg(1)\in F^{-(n-1)}\{g(1)\}}} \left| \varepsilon u_{g,\varepsilon} \,\mathrm{diam}(h(\mathcal{U}_\varepsilon(g(1)))) - \varepsilon g(1)\,|h'(g(1))| \right| + \frac{\log n}{n^2}$$

$$\le \left(\varepsilon g(1) C_g + |\varepsilon u_{g,\varepsilon} - g(1)| \right) \frac{1}{\varepsilon u_{g,\varepsilon}} u_{g,\varepsilon} \log n \sum_{\substack{f\in\Phi:\\ f(1)\in F^{-(n-1)}\{g(1)\}}} |f'(1)|$$

$$\xrightarrow{n\to\infty} \left(\varepsilon g(1) C_g + |\varepsilon u_{g,\varepsilon} - g(1)| \right) \frac{1}{g(1)},$$

where the convergence follows from Proposition 5.6.2 and (5.3). This inequality holds for all $x \in [0, 1]$, $n \in \mathbb{N}$ and the right-hand side vanishes for $\varepsilon \to 0$. Hence, using the condition for weak convergence in terms of distribution functions, we obtain that

$$\text{w-lim}_{n\to\infty} \rho_{g,n} = \lambda.$$

The proof of Theorem 5.6.1 now follows, if we use in the definition of $\rho_{g,n}$ the fact that $g(1)$ can be written in the form of a reduced fraction v/w and that then $|g'(1)| = w^{-2}$, as well as similarly, that $f(1)$ can be written in the form of a reduced fraction p/q and that then $|f'(1)| = q^{-2}$ (cf. Exercise 5.7.8). □

Finally, let us recall here the even Stern–Brocot sequence which was first defined in Section 1.3.1. For each $n \geq 0$, the n-th member of the Stern–Brocot sequence is denoted by $\mathcal{B}_n$ and the n-th member of the even Stern–Brocot sequence is defined to be $\mathcal{S}_n := \mathcal{B}_n \setminus \mathcal{B}_{n-1}$. Recalling that $\mathcal{S}_n$ is exactly the set $F^{-n}(\{1/2\})$, we obtain the following immediate corollary.

Corollary 5.6.4. *For the even Stern–Brocot sequence we have that*

$$\text{w-lim}_{n\to\infty} \log(n^2) \sum_{p/q\in\mathcal{S}_n} q^{-2} \delta_{p/q} = \lambda. \tag{5.4}$$

5.7 Exercises

Exercise 5.7.1. Show that any function of bounded variation f can be written as the difference $g - h$ of two bounded functions g and h, which are either both non-decreasing or both non-increasing.

Exercise 5.7.2. Show that if $p_i(x) := (1 + x)/((i + x)(i + 1 + x))$ and $f \in L^1(\lambda)$, then

$$\widehat{G}f(x) = \sum_{i=1}^{\infty} p_i(x) f\left(\frac{1}{i+x}\right).$$

Exercise 5.7.3. Let $n \in \mathbb{N}$ and denote $C := C(x_1, \ldots, x_n)$. Show that $P_G^n(\mathbb{1}_C)$ is equal to the derivative of the inverse branch of G^n that maps the unit interval onto the cylinder set C.

Exercise 5.7.4. Prove the case $n = 0$, $C := [0, 1]$ of Lemma 5.2.5.

Exercise 5.7.5. Let

$$GL_2(\mathbb{Z}) := \left\{ \begin{pmatrix} a & b \\ c & d \end{pmatrix} : a, b, c, d \in \mathbb{Z}, |ad - bc| = 1 \right\}$$

denote the group of invertible 2×2 matrices over the integers and let $\mathrm{Aut}\left(\widehat{\mathbb{C}}\right)$ denote the group of automorphism of $\widehat{\mathbb{C}}$ that is the set of bi-holomorphic mappings of $\widehat{\mathbb{C}}$. Show that the map

$$\mathrm{GL}_2\left(\mathbb{Z}\right) \to \mathrm{Aut}\left(\widehat{\mathbb{C}}\right), \begin{pmatrix} a & b \\ c & d \end{pmatrix} \mapsto \left(z \mapsto \frac{az+b}{cz+d}\right)$$

defines a group homomorphism and determine its kernel. Further, show that for every element $\varphi : z \mapsto \frac{az+b}{cz+d}$ in the image of this homomorphism we have $\left|\varphi'(z)\right| = |cz+d|^{-2}$.

Exercise 5.7.6. The function $t \mapsto \int \exp(t \cdot x)\,\mathrm{d}\mu(x)$ is called the moment generating function of the probability distribution μ. Show that the k-th derivative of this function in 0 can be used to find the k-th moment $M_k(\mu) := \int x^k\,\mathrm{d}\mu(x)$, $k \in \mathbb{N}_0$, of μ, if it exists.

Exercise 5.7.7. For each $g \in \Phi$ there exists a constant Δ_g such that all $\varepsilon > 0$ sufficiently small and for all $h \in \Phi$, we have

$$\left|\mathrm{diam}(h(\mathcal{U}_\varepsilon(g(1)))) - \varepsilon\left|(h'(g(1)))\right|\right| \leq \varepsilon^2\left|(hg)'(1)\right|\Delta_g.$$

Exercise 5.7.8. Show that for each $p/q \in (0,1)$ such that $\gcd(p,q) = 1$, there exists a unique element $f \in \Phi$ with $f(1) = p/q$, and we have $f'(1) = q^2$.

Exercise 5.7.9. Prove that for all $a, b \in (0,1)$ with $a < b$ we have

$$\lim_{n \to \infty} \log(n)\lambda\left(F^{-n}([a,b])\right) = \log(b/a).$$

Bibliography

[Aar81]   J. Aaronson. The asymptotic distributional behaviour of transformations preserving infinite measures. *J. Anal. Math.*, 39:203–234, 1981.

[Aar97]   J. Aaronson. *An introduction to infinite ergodic theory*, volume 50 of *Mathematical Surveys and Monographs*. American Mathematical Society, Providence, RI, 1997.

[Adl98]   R. L. Adler. Symbolic dynamics and Markov partitions. *Bull. Amer. Math. Soc. (N.S.)*, 35(1):1–56, 1998.

[BBDK96]   J. Barrionuevo, R. M. Burton, K. Dajani, and C. Kraaikamp. Ergodic properties of generalized Lüroth series. *Acta Arith.*, 74(4):311–327, 1996.

[Ber12a]   F. Bernstein. Über eine Anwendung der Mengenlehre auf ein aus der Theorie der säkularen Störungen herrührendes Problem. *Math. Ann.*, 71:417–439, 1912.

[Ber12b]   F. Bernstein. Über geometrische Wahrscheinlichkeit und über das Axiom der beschränkten Arithmetisierbarkeit der Beobachtungen. *Math. Ann.*, 72(4):585–587, 1912.

[Bir31]   G. D. Birkhoff. Proof of the ergodic theorem. *Proc. Natl. Acad. Sci. USA*, 17:656–660, 1931.

[Bor09]   E. Borel. Les probabilités denombrables et leurs applications arithmétiques. *Rend. Circ. Mat. Palermo*, 27:247–271, 1909.

[Bro61]   A. Brocot. Calcul des rouages par approximation, nouvelle méthode. *Revue chronométrique*, 3:186–194, 1861.

[CO60]   R. V. Chacon and D. S. Ornstein. A general ergodic theorem. *Illinois J. Math.*, 4:153–160, 1960.

[Coh80]   D. L. Cohn. *Measure theory*. Birkhäuser, Boston, Mass., 1980.

[Den38]   A. Denjoy. Sur une fonction réelle de Minkowski. *J. Math. Pures Appl. (9)*, 17:105–151, 1938.

[DK96]   K. Dajani and C. Kraaikamp. On approximation by Lüroth series. *J. Théor. Nombres Bordeaux*, 8(2):331–346, 1996.

[DK02]   K. Dajani and C. Kraaikamp. *Ergodic theory of numbers*, volume 29 of *Carus Mathematical Monographs*. Mathematical Association of America, Washington, DC, 2002.

[DS88]   N. Dunford and J. T. Schwartz. *Linear operators. Part I*. Wiley Classics Library. John Wiley & Sons, Inc., New York, 1988. General theory, With the assistance of William G. Bade and Robert G. Bartle, Reprint of the 1958 original, A Wiley-Interscience Publication.

[Dud89]   R. M. Dudley. *Real analysis and probability*. The Wadsworth & Brooks/Cole Mathematics Series. Wadsworth & Brooks/Cole Advanced Books & Software, Pacific Grove, CA, 1989.

[DV86]   H. G. Diamond and J. D. Vaaler. Estimates for partial sums of continued fraction partial quotients. *Pacific J. Math.*, 122(1):73–82, 1986.

[EFP49]   P. Erdös, W. Feller, and H. Pollard. A property of power series with positive coefficients. *Bull. Amer. Math. Soc.*, 55:201–204, 1949.

[Eri70]   K. B. Erickson. Strong renewal theorems with infinite mean. *Trans. Amer. Math. Soc.*, 151:263–291, 1970.

[EW11]   M. Einsiedler and T. Ward. *Ergodic theory with a view towards number theory*, volume 259 of *Graduate Texts in Mathematics*. Springer-Verlag London, Ltd., London, 2011.

[Fal14]   K. Falconer. *Fractal geometry*. John Wiley & Sons, Ltd., Chichester, third edition, 2014. Mathematical foundations and applications.

[Far16]   J. Farey. On a curious property of vulgar fractions. *Phil. Mag. Ser. 1*, 47(217):385–386, 1816.

[Fel68a]   W. Feller. *An introduction to probability theory and its applications. Vol. I*. Third edition. John Wiley & Sons, Inc., New York-London-Sydney, 1968.

[Fel68b]   W. Feller. *An introduction to probability theory and its applications. Vol. II*. Third edition. John Wiley & Sons, Inc., New York-London-Sydney, 1968.

[FK10]   J. Fiala and P. Kleban. Intervals between Farey fractions in the limit of infinite level. *Ann. Sci. Math. Québec*, 34(1):63–71, 2010.

[Gal72]   J. Galambos. Some remarks on the Lüroth expansion. *Czechoslovak Math. J.*, 22(97):266–271, 1972.

[Gal73]   J. Galambos. The largest coefficient in continued fractions and related problems. In *Diophantine approximation and its applications (Proc. Conf., Washington, D.C., 1972)*, pages 101–109. Academic Press, New York, 1973.

[Gam01]   T. W. Gamelin. *Complex analysis*. Undergraduate Texts in Mathematics. Springer-Verlag, New York, 2001.

[Gan01]   C. Ganatsiou. On some properties of the Lüroth-type alternating series representations for real numbers. *Int. J. Math. Math. Sci.*, 28(6):367–373, 2001.

[GL63]   A. Garsia and J. Lamperti. A discrete renewal theorem with infinite mean. *Comment. Math. Helv.*, 37:221–234, 1962/1963.

[GLJ93]   Y. Guivarc'h and Y. Le Jan. Asymptotic winding of the geodesic flow on modular surfaces and continued fractions. *Ann. Sci. École Norm. Sup. (4)*, 26(1):23–50, 1993.

[GLJ96]   Y. Guivarch and Y. Le Jan. Note rectificative: "Asymptotic winding of the geodesic flow on modular surfaces and continued fractions" (*Ann. Sci. École Norm. Sup. (4)* 26(1):23–50, 1993; MR1209912 (94a:58157)). *Ann. Sci. École Norm. Sup. (4)*, 29(6):811–814, 1996.

[GM88]   M. C. Gutzwiller and B. B. Mandelbrot. Invariant multifractal measures in chaotic Hamiltonian systems, and related structures. *Phys. Rev. Lett.*, 60(8):673–676, 1988.

[Gou04]   S. Gouëzel. Sharp polynomial estimates for the decay of correlations. *Israel J. Math.*, 139:29–65, 2004.

[Gou11]   S. Gouëzel. Correlation asymptotics from large deviations in dynamical systems with infinite measure. *Colloq. Math.*, 125(2):193–212, 2011.

[Gut11]   S. B. Guthery. *A motif of mathematics*. Docent Press, Boston, MA, 2011. History and application of the mediant and the Farey sequence.

[Gut13]   A. Gut. *Probability: A Graduate Course*. Springer Texts in Statistics. Springer, New York, second edition, 2013.

[Hal56]   P. R. Halmos. *Lectures on Ergodic Theory*. Publications of the Mathematical Society of Japan, no. 3. The Mathematical Society of Japan, 1956.

[Hee15]   B. Heersink. An effective estimate for the Lebesgue measure of preimages of iterates of the farey map. Advances in Mathematics Volume 291, 19 March 2016, Pages 621–634.

[Hei87]   L. Heinrich. Rates of convergence in stable limit theorems for sums of exponentially ψ-mixing random variables with an application to metric theory of continued fractions. *Math. Nachr.*, 131:149–165, 1987.

[Hen00]   D. Hensley. The statistics of the continued fraction digit sum. *Pacific J. Math.*, 192(1):103–120, 2000.

[HW08]   G. H. Hardy and E. M. Wright. *An introduction to the theory of numbers*. Oxford University Press, Oxford, sixth edition, 2008. Revised by D. R. Heath-Brown and J. H. Silverman, With a foreword by Andrew Wiles.

[Ios92]   M. Iosifescu. A very simple proof of a generalization of the Gauss-Kuzmin-Lévy theorem on continued fractions, and related questions. *Rev. Roumaine Math. Pures Appl.*, 37(10):901–914, 1992.

[Iso11]   S. Isola. From infinite ergodic theory to number theory (and possibly back). *Chaos, Solitons and Fractals*, 44(7):467–479, 2011.

[Jar29]   V. Jarník. Zur metrischen Theorie der diophantischen Approximationen. Przyczynek do metrycznej teorji przyblizeń diofantowych. *Prace Mat.-Fiz.*, 36:91–106, 1929.

[JKS13]   J. Jaerisch, M. Kesseböhmer, and B. O. Stratmann. A Fréchet law and an Erdős-Philipp law for maximal cuspidal windings. *Ergodic Theory Dynam. Syst.*, 33(4):1008–1028, 2013.

[Kak43]   S. Kakutani. Induced measure preserving transformations. *Proc. Imp. Acad. Tokyo*, 19:635–641, 1943.

[Khi35]   A. Khintchine. Metrische Kettenbruchprobleme. *Compositio Math.*, 1:361–382, 1935.

[Khi64]   A. Ya. Khinchin. *Continued fractions*. The University of Chicago Press, Chicago, Ill.-London, 1964.

[KKK91]   S. Kalpazidou, A. Knopfmacher, and J. Knopfmacher. Metric properties of alternating Lüroth series. *Portugal. Math.*, 48(3):319–325, 1991.

[KKS15]   J. Kautzsch, M. Kesseböhmer, and T. Samuel. On the convergence to equilibrium of unbounded observables under a family of intermittent interval maps. *Ann. Henri Poincaré*, 17(9):2585–2621, 2016.

[KKSS15]   J. Kautzsch, M. Kesseböhmer, T. Samuel, and B. O. Stratmann. On the asymptotics of the α-Farey transfer operator. *Nonlinearity*, 28(1):143–166, 2015.

[KMS12]   M. Kesseböhmer, S. Munday, and B. O. Stratmann. Strong renewal theorems and Lyapunov spectra for α-Farey and α-Lüroth systems. *Ergodic Theory Dynam. Syst.*, 32(3):989–1017, 2012.

[Koo31]   B. O. Koopman. Hamiltonian systems and transformation in hilbert space. *Proc. Natl. Acad. Sci.*, 17(5):315–318, 1931.

[KS07]   M. Kesseböhmer and B. O. Stratmann. A multifractal analysis for Stern–Brocot intervals, continued fractions and Diophantine growth rates. *J. Reine Angew. Math.*, 605:133–163, 2007.

[KS08a]   M. Kesseböhmer and M. Slassi. Large deviation asymptotics for continued fraction expansions. *Stoch. Dyn.*, 8(1):103–113, 2008.

[KS08b]   M. Kesseböhmer and B. O. Stratmann. Fractal analysis for sets of non-differentiability of Minkowski's question mark function. *J. Number Theory*, 128(9):2663–2686, 2008.

[KS12a]   M. Kesseböhmer and B. O. Stratmann. A dichotomy between uniform distributions of the Stern–Brocot and the Farey sequence. *Unif. Distrib. Theory*, 7(2):21–33, 2012.

[KS12b]   M. Kesseböhmer and B. O. Stratmann. On the asymptotic behaviour of the Lebesgue measure of sum-level sets for continued fractions. *Discrete Contin. Dyn. Syst.*, 32(7):2437–2451, 2012.

[Kuz28]   R. O. Kuz'min. Sur un problème de Gauss. *Anni Congr. Intern. Bologne*, 6:83–89, 1928.

[Len12]   M. Lenci. Infinite-volume mixing for dynamical systems preserving an infinite measure. *Procedia IUTAM*, 5:204–219, 2012. IUTAM Symposium on 50 Years of Chaos: Applied and Theoretical.

[Len13]   M. Lenci. Exactness, K-property and infinite mixing. *Publ. Mat. Urug.*, 14:159–170, 2013.

[Len14]   M. Lenci. Uniformly expanding Markov maps of the real line: exactness and infinite mixing. *preprint: arXiv:1404.2212*, 2014.

[Lev29]   P. Lévy. Sur les lois de probabilité dont dependent les quotients complets et incomplets d'une fraction continue. *Bull. Soc. Math. France*, 57:178–194, 1929.

[Lév52]   P. Lévy. Fractions continues aléatoires. *Rend. Circ. Mat. Palermo (2)*, 1:170–208, 1952.

[Lin71]   M. Lin. Mixing for Markov operators. *Z. Wahrscheinlichkeitstheorie und Verw. Gebiete*, 19:231–242, 1971.

[LM95]   D. Lind and B. Marcus. *An introduction to symbolic dynamics and coding*. Cambridge University Press, Cambridge, 1995.

[Lür83]   J. Lüroth. Ueber eine eindeutige Entwickelung von Zahlen in eine unendliche Reihe. *Math. Ann.*, 21(3):411–423, 1883.

[Mar92]   G. Markowsky. Misconceptions about the golden ratio. *Coll. Math. J.*, 23(1):2–19, 1992.

[Min10]   H. Minkowski. Geometrie der Zahlen. In 2 Lieferungen. II. (Schluß-) Lieferung. Leipzig: B. G. Teubner. VIII + S. 241–256 (1910), 1910.

[MN13]   T. Miernowski and A. Nogueira. Exactness of the Euclidean algorithm and of the Rauzy induction on the space of interval exchange transformations. *Ergodic Theory Dynam. Syst.*, 33(1):221–246, 2013.

[MT12]   I. Melbourne and D. Terhesiu. Operator renewal theory and mixing rates for dynamical systems with infinite measure. *Invent. Math.*, 189(1):61–110, 2012.

[MT15]   I. Melbourne and D. Terhesiu. Erratum to: Operator renewal theory and mixing rates for dynamical systems with infinite measure. *Invent. Math.*, 202(3):1269–1272, 2015.

[MU03]   R. D. Mauldin and M. Urbański. *Graph directed Markov systems*, volume 148 of *Cambridge Tracts in Mathematics*. Cambridge University Press, Cambridge, 2003. Geometry and dynamics of limit sets.

[Mun11]   S. Munday. *Finite and infinite ergodic theory for linear and conformal dynamical systems*. PhD thesis, University of St. Andrews, 2011.

[Mun14]   S. Munday. On the derivative of the α-Farey-Minkowski function. *Discrete Contin. Dyn. Syst.*, 34(2):709–732, 2014.

[Neu32]   J. von Neumann. Proof of the quasi-ergodic hypothesis. *Proc. Natl. Acad. Sci. USA*, 18:70–82, 1932.

[Par81]   W. Parry. *Topics in ergodic theory*, volume 75 of *Cambridge Tracts in Mathematics*. Cambridge University Press, Cambridge–New York, 1981.

[Phi88]   W. Philipp. Limit theorems for sums of partial quotients of continued fractions. *Monatsh. Math.*, 105(3):195–206, 1988.

[Phi76]   W. Philipp. A conjecture of Erdős on continued fractions. *Acta Arith.*, 28(4):379–386, 1975/76.

[Roh48]   V. Rohlin. A "general" measure-preserving transformation is not mixing. *Doklady Akad. Nauk SSSR (N.S.)*, 60:349–351, 1948.

[Rok64]   V. A. Rokhlin. Exact endomorphisms of a Lebesgue space. *Transl., Ser. 2, Am. Math. Soc.*, 39:1–36, 1964.

[RS92]   A. M. Rockett and P. Szüsz. *Continued fractions*. World Scientific Publishing Co., Inc., River Edge, NJ, 1992.

[Rud87]   W. Rudin. *Real and complex analysis*. McGraw-Hill Book Co., New York, third edition, 1987.

[Rud91]   W. Rudin. *Functional analysis*. *International Series in Pure and Applied Mathematics*. McGraw-Hill, Inc., New York, second edition, 1991.

[Sal43]   R. Salem. On some singular monotonic functions which are strictly increasing. *Trans. Amer. Math. Soc.*, 53:427–439, 1943.

[Šal68]   T. Šalát. Zur metrischen Theorie der Lürothschen Entwicklungen der reellen Zahlen. *Czechoslovak Math. J.*, 18(93):489–522, 1968.

[Sar02]   O. Sarig. Subexponential decay of correlations. *Invent. Math.*, 150(3):629–653, 2002.

[Sch95]   F. Schweiger. *Ergodic theory of fibred systems and metric number theory*. Oxford Science Publications. The Clarendon Press, Oxford University Press, New York, 1995.

[Sen76]   E. Seneta. *Regularly varying functions*, volume 508 of *Lecture Notes in Mathematics*. Springer-Verlag, Berlin–New York, 1976.

[Ste58]   M. A. Stern. Ueber eine zahlentheoretische Funktion. *J. Reine Angew. Math.*, 55:193–220, 1858.

[SW07]   L.-M. Shen and J. Wu. On the error-sum function of Lüroth series. *J. Math. Anal. Appl.*, 329(2):1440–1445, 2007.

[Wal82]   P. Walters. *An introduction to ergodic theory*, volume 79 of *Graduate Texts in Mathematics*. Springer-Verlag, New York–Berlin, 1982.

[Wir74]   E. Wirsing. On the theorem of Gauss–Kusmin–Lévy and a Frobenius-type theorem for function spaces. *Acta Arith.*, 24:507–528, 1973/74. Collection of articles dedicated to Carl Ludwig Siegel on the occasion of his seventy-fifth birthday, V.

[WX11]   S. Wang and J. Xu. On the Lebesgue measure of sum-level sets for Lüroth expansion. *J. Math. Anal. Appl.*, 374(1):197–200, 2011.

[Zwe04]   R. Zweimüller. Hopf's ratio ergodic theorem by inducing. *Colloq. Math.*, 101(2):289–292, 2004.

Index

www.ingramcontent.com/pod-product-compliance
Lightning Source LLC
Chambersburg PA
CBHW081523220326
41598CB00036B/6306